Volume 65

Genetic Dissection of Neural Circuits and Behavior

Advances in Genetics, Volume 65

Serial Editors

Jay C. Dunlap
Dartmouth Medical School, Hanover, NH, USA

Theodore Friedmann
University of California at San Diego, School of Medicine, USA

Stephen F. Goodwin
University of Oxford, Oxford, UK

Volume 65

Genetic Dissection of Neural Circuits and Behavior

Edited by

Stephen F. Goodwin

Department of Physiology, Anatomy and Genetics
University of Oxford
Oxford, United Kingdom

AMSTERDAM • BOSTON • HEIDELBERG • LONDON
NEW YORK • OXFORD • PARIS • SAN DIEGO
SAN FRANCISCO • SINGAPORE • SYDNEY • TOKYO
Academic Press is an imprint of Elsevier

Academic Press is an imprint of Elsevier

525 B Street, Suite 1900, San Diego, CA 92101-4495, USA
30 Corporate Drive, Suite 400, Burlington, MA 01803, USA
32 Jamestown Road, London, NW1 7BY, UK
Radarweg 29, POBox 211, 1000 AE Amsterdam, The Netherlands

First edition 2009

ISBN: 978-0-12-374836-2
ISSN: 0065-2660

For information on all Academic Press publications
visit our website at elsevierdirect.com

Printed and bound in USA

09 10 11 12   10 9 8 7 6 5 4 3 2 1

Working together to grow
libraries in developing countries

www.elsevier.com | www.bookaid.org | www.sabre.org

ELSEVIER   BOOK AID
            International   Sabre Foundation

Contents

Contributors

Numbers in parentheses indicate the pages on which the authors' contributions begin.

Ted Abel (1) Department of Biology, University of Pennsylvania, Philadelphia, PA, USA

Jeremy Dittman (39) Department of Biochemistry, Weill Cornell Medical College, New York, NY, USA

Robbert Havekes (1) Department of Biology, University of Pennsylvania, Philadelphia, PA, USA

Stephen Nurrish (145) MRC Cell Biology Unit, MRC Laboratory for Molecular Cell Biology and Department of Neurobiology, Physiology and Pharmacology, University College London, London, United Kingdom

Borja Perez-Mansilla (145) MRC Cell Biology Unit, MRC Laboratory for Molecular Cell Biology and Department of Neurobiology, Physiology and Pharmacology, University College London, London, United Kingdom

Julie H. Simpson (79) HHMI Janelia Farm Research Campus, Asburn, VA, USA

1

Genetic Dissection of Neural Circuits and Behavior in *Mus musculus*

Robbert Havekes and Ted Abel

Department of Biology, University of Pennsylvania, Philadelphia, PA, USA

Advances in Genetics, Vol. 65
0065-2660/09 $35.00
DOI: 10.1016/S0065-2660(09)65001-X

ABSTRACT

One of the major challenges in the field of neurobiology is to elucidate the molecular machinery that underlies the formation and storage of memories. For many decades, genetic studies in the fruit fly (*Drosophila melanogaster*) have provided insight into the role of specific genes underlying memory storage. Although these pioneering studies were groundbreaking, a transition to a mammalian system more closely resembling the human brain is critical for the translation of basic research findings into therapeutic strategies in humans. Because the mouse (*Mus musculus*) shares the complex genomic and neuroanatomical organization of mammals and there is a wealth of molecular tools that are available to manipulate gene function in mice, the mouse has become the primary model for research into the genetic basis of mammalian memory. Another major advantage of mouse research is the ability to examine *in vivo* electrophysiological processes, such as synaptic plasticity and neuronal firing patterns during behavior (e.g., the analysis of place cell activity). The focus on mouse models for memory research has led to the development of sophisticated behavioral protocols capable of exploring the role of particular genes in distinct phases of learning and memory formation, which is one of the major accomplishments of the past decade. In this chapter, we will give an overview of several state of the art genetic approaches to study gene function in the mouse brain in a spatially and temporally restricted fashion.   © 2009, Elsevier Inc.

I. CONVENTIONAL GENE TARGETING STRATEGIES

In 2007, the Nobel Prize in physiology or medicine was awarded to Mario Capecchi, Martin Evans, and Oliver Smithies for the development of specific gene modification techniques and stem cell technology that led to the creation of the first "knockout mouse." To date, thousands of knockout mouse models have been developed that have revolutionized our understanding of the molecular mechanisms responsible for various biological processes, including the signaling pathways underlying memory storage. Recently, the mouse knockout project (KOMP, see http://www.knockoutmouse.org), based on reverse genetics (from gene to phenotype), was established to produce and phenotype knockouts for all mouse genes. In addition to knockout mouse models, knock-in mouse models have been used as an alternative approach to determine the function of specific functional domains within proteins of interest. Knockout models are based on the deletion of an entire gene leading to complete inactivation, but knock-in models are based on a subtle alteration of a targeted gene sequence, for example, by introducing one or more point mutations to alter specific functional domains of proteins rather than deleting the entire protein. One example of the successful use of a knock-in strategy, was the recent discovery that a single transcription factor-binding domain of the CREB-binding protein (CBP), a transcriptional coactivator, is pivotal for the formation of long-term memories (Wood et al., 2005, 2006). Knock-in mice also enable researchers to model disease-causing mutations found in human patients.

As an alternative to knockout and knock-in mutations, transgenic methods have been used for more than 25 years as a reverse genetic approach to study the role of particular gene products in biological processes. The first transgenic mouse was generated in 1982 by Palmiter and colleagues via injection of a metallothionein-growth hormone fusion gene directly into the pronucleus of a fertilized single cell mouse embryo. This fusion gene incorporated into the genome and was subsequently expressed in essentially every cell in the embryo throughout development and adulthood (Palmiter et al., 1982). Such a ubiquitous expression pattern could make the functional interpretation of specific gene functions difficult due to various confounding side effects even including premature death. For example, the role of the NR1 subunit of the N-methyl-D-aspartate receptor (NMDAr) in learning and memory could not be unraveled using a classical knockout approach because mice lacking the NR1 subunit of the NMDAr die as neonates (Forrest et al., 1994). In another example, initial studies on mice with a mutation in the tyrosine kinase gene *fyn* revealed that loss of the *fyn* gene resulted in deficits in long-term potentiation (LTP, the persistent increase in synaptic strength following high-frequency stimulation of a chemical synapse) and spatial learning (Grant et al., 1992). However, *fyn* deletion also caused developmental abnormalities in various brain structures, for example, an increased number of cells in hippocampus area CA3 and the dentate gyrus (DG). Similarly, studies with mice exhibiting a mutant form of the α-calcium-calmodulin-dependent kinase II (α-CaMKII), a synaptic protein that is strongly enriched in forebrain neurons including those in the hippocampus, indicated that loss of α-CaMKII-function resulted in impaired spatial memory and LTP (Silva et al., 1992a,b). However, simultaneous studies indicated that loss of functional α-CaMKII resulted in behavioral abnormalities unrelated to memory formation including a decreased fear response and increased aggression (Chen et al., 1994). Another risk with conventional gene knockout is that loss of a particular gene may result in compensation by other related genes. For example, loss of the alpha and delta isoform of the cAMP response element binding protein (CREB) results in a compensatory upregulation of the beta isoform of CREB (Blendy et al., 1996). This beta isoform is expressed in low levels in wild-type mice, but is dramatically upregulated in mutant mice. The fact that beta CREB is expressed in several but not all regions where CREB function has been implicated and that the specific phenotype observed in these mutants was different from the phenotype observed in mice expressing a dominant-negative mutant form of CREB (Struthers et al., 1991), suggest that the beta isoform at least partly compensates for the loss of the alpha and delta isoforms (Blendy *et al.*, 1996; Gass *et al.*, 1998). Thus, although the initial approach to study the role of particular genes were groundbreaking, the interpretation of the phenotypes of these first mutant mice remain difficult not just for the molecular reasons described here, but also for difficulties in interpreting the limited behavioral experiments that were carried out in the initial studies (Deutsch, 1993).

To circumvent these problems, two major improvements in experimental approach were needed. The first was to avoid the occurrence of developmental and nonspecific abnormalities in mice with permanent and global gene manipulations. Thus, more sophisticated molecular approaches were needed to allow researchers to affect gene function in a temporally and spatially restricted fashion, as opposed to constitutively and globally. The second major issue was the lack of sensitive behavioral paradigms that could distinguish between learning-specific and learning-unrelated phenotypes. In human patients with brain lesions, this is done by using "double dissociation approaches." For example, Bechara and colleagues (1995) showed that a patient with bilateral damage to the amygdala failed to acquire conditioned autonomic responses to visual or auditory stimuli, although the patient could learn which visual or auditory stimuli were paired with the unconditioned stimulus (indicative for a normal declarative memory). In contrast, a patient with bilateral hippocampal lesions was impaired in the formation of declarative memories but acquired the autonomic conditioning in the same task. In mice, contextual and tone-cued fear conditioning are used in parallel to suggest whether the manipulation of a specific gene affects either hippocampal function or amygdala function. Contextual fear conditioning relies on both the hippocampus and amygdala. The hippocampus is required to make an association between the context (e.g., the shock box, the conditioned stimulus) and a negative stimulus (e.g., a mild electrical shock, the unconditioned stimulus). If the mice make this association, then they will show freezing behavior when reexposed to the same context. Freezing behavior, defined as the lack of movement except for respiration, is initiated by the context-shock association and mediated by the amygdala. Tone-cued fear conditioning is a paradigm in which mice make an association between a tone, that is presented during training and a mild foot shock (LeDoux, 2000). In contrast to contextual fear conditioning, tone-cued fear conditioning does not require the hippocampus. Using both behavioral paradigms, Abel *et al.* (1997) showed that mice with reduced activity of the cAMP dependent protein kinase (PKA) in forebrain neurons were impaired in contextual fear conditioning but not in tone-cued fear conditioning, indicating that memory defects in this mouse line are likely due to defects in hippocampal PKA function. Although tone-cued fear conditioning in combination with contextual fear conditioning provides strong support for a hippocampal site of biochemical malfunction, this is not a true double dissociation as described by Bechara and colleagues (1995). This is particularly due to the lack of sophisticated behavioral paradigms and the inability to finely regulate spatial resolution of gene manipulation (e.g., hippocampus or amygdala specific manipulations).

In addition to the generation of mutant mice using reverse genetics, the use of forward genetics (from phenotype to gene) has been a successful approach to uncover genes that drive the molecular mechanism of specific behaviors. For instance, forward genetic screens using the point mutagen *N*-ethyl-*N*-nitrosourea have yielded great success in the discovery of genes that control certain behaviors. This approach has been especially successful in elucidating genes that control circadian rhythms, so-called "clock genes" (Takahashi *et al.*, 2008), because highly defined behavioral protocols exist to determine the activity profiles of mice with great precision and high throughput. Applying this approach to learning in mice has been more difficult, in part because of the behavioral complexity of learning tasks and because of the fact that animals cannot be repeatedly tested in learning paradigms (Reijmers *et al.*, 2006).

II. SPATIALLY RESTRICTED GENE MANIPULATION

To avoid side effects induced by overall gene manipulation and to achieve a better understanding of a particular gene function in a specific brain region, more sophisticated methods have been developed that result in spatially restricted manipulation of a gene. One powerful approach to locally manipulate gene function uses gene promoters to restrict transgene expression. For example, the use of promoters like Nestin (Cheng *et al.*, 2004), PrP (Fischer *et al.*, 1996), and neuron-specific enolase (Forss-Petter *et al.*, 1990) result in transgene expression throughout the brain but not in other parts of the body. For further restriction to specific brain regions or cell types, researchers used more specific promoters, including the α-CaMKII promoter for postnatal forebrain specific expression in excitatory neurons (Abel *et al.*, 1997; Mayford *et al.*, 1996), glial fibrillary acidic protein (GFAP) for astrocyte-specific expression (Brenner *et al.*, 1994; Casper *et al.*, 2007; Halassa *et al.*, 2009; Pascual *et al.*, 2005), proteolipid protein (PLP) oligodendrocyte-specific expression (Fuss *et al.*, 2001), and L7 for expression in cerebellar Purkinje cells and retinal bipolar neurons (Oberdick *et al.*, 1990). Large-scale screening projects have been initiated to determine the brain expression patterns of a tremendous number of specific promoters, allowing researchers to choose a promoter that targets a particular class of neurons, such as those that produce a specific neurotransmitter, or express a specific receptor. Examples of these screening projects are GENSAT (http://www.gensat.org) and the Allen Brain Atlas (http://www.brain-map.org). Although the transgenic approach allows for region-specific overexpression of a gene, it is possible that information from transgenic mouse lines may not translate into the requirement for a mutated gene or pathway in memory. Interfering with gene function through transgenesis is necessarily through overexpression of a protein, which can potentially interfere with genes other than the intended targets. Therefore, other methods have been developed that delete single genes in a tissue or cell-type specific manner as described below.

A. The Cre/loxP system

In contrast to transgenic methods that spatially restrict overexpression of an interfering transgene, the Cre/loxP system allows for the knockout of a single gene in a cell-type or site-specific manner. The Cre/loxP system is based on a phage P1-derived site-specific recombination system (Sauer and Henderson, 1988). This system is based on the ability of Cre recombinase to induce the recombination of 34 bp loxP recognition sequences. The loxP sites can be inserted into the genome of embryonic stem cells using homologous recombination, flanking one or more exons of the target gene (referred to as a "floxed" gene). Mice homozygous for the floxed gene are then crossed to a transgenic mouse line that expresses Cre recombinase under control of a spatially restricted promoter. In offspring that are homozygous for the floxed gene and carry the Cre transgene, the floxed gene will be excised. Depending on the promoter used to drive Cre expression, the loss of the floxed gene can be more or less restricted. Gu *et al.* (1994) generated the first inducible KO mice using this method by generating a T-cell specific knockout mouse for the DNA polymerase beta gene, thus showing proof of principle.

B. Investigating NMDA receptor function in specific subregions of the brain using the Cre/loxP system

The NMDAr is an ionotropic glutamate receptor that allows flow of calcium ions (Ca^{2+}) into the cell. Gating of the channel is voltage dependent and can be controlled by binding its ligand glutamate. The influx of Ca^{2+} into neurons plays a pivotal role in synaptic plasticity thought to underlie learning and memory formation. The first indication that NMDArs could play a pivotal role in the processes underlying memory storage came from a study by Collingridge and colleagues, demonstrating that NMDArs are required for the persistent change in strength of synaptic connections, a process referred to as LTP (Collingridge *et al.*, 1983). Activity-dependent changes in synaptic strength is a form of synaptic plasticity that is a model for certain types of long-term memory (Bliss and Collingridge, 1993; Hebb, 1949; Martin *et al.*, 2000). LTP in the CA1 region of the hippocampus has distinct temporal phases (Nguyen and Woo, 2003): an early transient phase (E-LTP) and a late persistent phase (L-LTP) that lasts for up to 8 h in hippocampal slices (Frey *et al.*, 1993), and for days in the intact animal (Abraham *et al.*, 1993). Both E-LTP and L-LTP in the CA1 region of the hippocampus depend on NMDArs and on the activation of kinases such as CaMKII, whereas L-LTP (but not E-LTP) shares with long-term memory, a requirement for transcription and translation as well as cAMP and PKA activity (Abel *et al.*, 1997; Duffy and Nguyen, 2003; Frey *et al.*, 1996; Scharf *et al.*, 2002; Woo *et al.*, 2000).

Three years after the study by Collingridge in 1983, Morris *et al.* (1986) showed that blocking NMDAr activity using the drug (2*R*)-amino-5-phospho-novaleric acid (AP5) prevents the formation of memories for the location of a submerged platform in the Morris water maze, based on visual cues in the room. In addition, the authors showed that application of AP5 impairs hippocampal LTP *in vivo* demonstrating the causal role of the NMDA receptor in hippocampus-dependent memory formation. Although these pharmacological studies indicate a pivotal role of the hippocampus in the formation of long-term memories, it could not be excluded that AP5 in the study by Morris *et al.* also affected NMDAr function in regions other than the hippocampus that are also implicated in memory formation, such as the neocortex (Morris *et al.*, 1986).

To analyze the role of the NMDAr specifically in the hippocampus, a genetic approach was needed that could tightly control the loss of NMDA receptor function to a specific cell type or group of cells. The discovery of the gene that encoded the NR1 subunit, a crucial subunit of the NMDA receptor (Moriyoshi *et al.*, 1991; Nakanishi, 1992) enabled researchers to manipulate NMDAr function genetically in a spatially restricted manner, because loss of this single receptor subunit prevents the expression of functional NMDArs.

The Tonegawa laboratory successfully applied the Cre/loxP system to study the NMDAr in distinct hippocampal subregions in learning and memory formation. They first generated several mouse lines that expressed Cre recombi-nase under control of the α-CamKII promoter to drive the expression of Cre recombinase specifically in neurons of the forebrain postnatally (Fig. 1.1A, Tsien *et al.*, 1996a). Several of the different founder lines showed variation in the expression pattern of Cre recombinase 3 weeks postnatally, despite the fact that the same promoter (α-CaMKII) was used to drive Cre expression (Tsien *et al.*, 1996a). This variation in Cre expression is likely due to distinct sites of transgene insertion in the genome. Some founders expressed Cre recombinase in all forebrain regions, including the three major hippocampal regions (e.g., areas CA1, CA3, and the DG), striatum and cortex, but other lines had a more restricted expression pattern. One of the founder lines was of particular interest because Cre recombinase expression was restricted to area CA1 of the hippo-campus at the age of 3 weeks. This very narrow Cre expression allowed the Tonegawa laboratory to examine the role of the NMDAr specifically within the hippocampal CA1 area for learning and memory formation as well as synaptic plasticity.

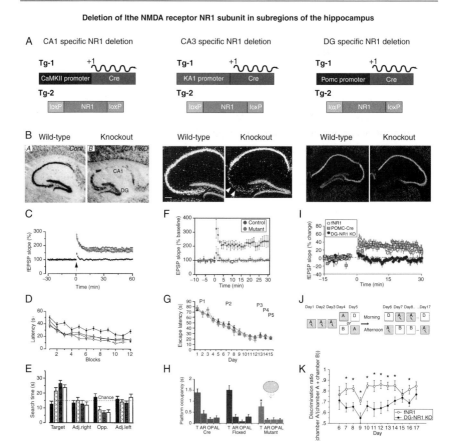

Figure 1.1. The deletion of NMDA receptor NR1 subunit in specific subregions of the hippocampus. (A) Loss of NMDA receptor function in specific hippocampal areas can be achieved by using specific promoters to drive Cre recombinase. (B) *In situ* hybridization of *NMDAR1* mRNA from wild types (left panels) and subregion specific NR1 knockout mice (right panels). (C) CA1-restricted deletion of the NR1 subunit impairs LTP in the Schaffer collateral pathway (wild-types open symbols, mutant closed circles), (D) CA1-specific knockout mice are impaired during training in the spatial version of the Morris water maze (wild-types open symbols, mutant closed circles), and (E) during the probe test (black bar indicates mutants). (F) CA3 specific deletion of the NR1 subunit blocks LTP at the recurrent commissural/associational synapses, (G) does not affect training in the spatial version of the Morris water maze, (H) but results in a reduced preference for the target quadrant during a probe trial with partial cues, indicating an impairment in pattern completion. (I) Perforant path LTP is lost as a consequence of ablation of the NR1 subunit in the dentate gyrus. (J) A design to test context discrimination in wild type and dentate gyrus-specific NMDA receptor knockout mice. For the first 3 days of conditioning, mice visited only chamber A and each day received a single foot shock. Freezing was measured once in chamber A and once in chamber B over the subsequent

By crossing this CA1-specific Cre expressing line with a floxed NR1 line, Tsien and colleagues (1996b) generated a mutant mouse in which the NR1 subunit was lost specifically in the CA1 region of the hippocampus at 3 weeks of age (Fig. 1.1B). These mice lacked NMDAr-mediated synaptic currents and LTP at CA1 synapses (Fig. 1.1C). Behavioral studies showed that loss of NMDAr function caused difficulties with learning and remembering the location of a submerged platform in the Morris water maze using distal cues (Fig. 1.1D and E). The mutants were not impaired in a nonspatial version of the same task that does not require the hippocampus, indicating that specifically hippocampus-dependent memory processing was affected by loss of NMDArs in area CA1 of the hippocampus (Tsien *et al.*, 1996b). In a study using the same mutant mice, McHugh *et al.* (1996) determined whether loss of CA1 NMDArs affected the properties of particular hippocampal cells called place cells: cells that exhibit a high rate of firing whenever an animal is in a specific location in an environment corresponding to the cell's "place field." Place cells were first described by O'Keefe and Dostrovsky (1971) and are thought to be important for spatial mapping of novel environments and thus pivotal for the formation of long-term spatial memories. They found that mice lacking NMDArs in area CA1 of the hippocampus had reduced spatial specificity of individual place fields (e.g., place cells activity was more widely distributed throughout the environment). Thus, their study indicated that CA1 NMDAr-mediated synaptic plasticity is necessary for the generation of a proper representation of space.

Using a similar approach, the Tonegawa laboratory next focused on area CA3 of the hippocampus, an area that is one of the main input sites for area CA1. To restrict the loss of the NMDAr to area CA3, Nakazawa and colleagues (2002) utilized a transgenic mouse line in which Cre expression was driven by the KA1 receptor (Fig. 1.1A). Expression from this promoter is specific for area CA3 and the DG in adult mice (Kask *et al.*, 2000). In mice that were homozygous for the floxed NR1 subunit and positive for the Cre transgene, loss of the NR1 subunit was restricted to area CA3 of the hippocampus (Nakazawa *et al.*, 2002, 2003). Electrophysiological studies on these mice revealed that loss of the NR1 subunit in area CA3 impairs LTP in the recurrent collaterals (input from CA3 neurons to other CA3 neurons which might serve as an auto associative network) (Fig. 1.1F). As a consequence of the loss of NMDAr in area CA3,

---

2 days. During days 6–17, mice visited each chamber daily (receiving a shock in one of the two), and freezing was assessed during the first 3 min in each chamber. (K) DG-restricted ablation of NMDA receptors results in impaired pattern separation indicated by a reduced discrimination ratio between a shocked context (context A) and a non-shocked context (context B). *Sources:* McHugh *et al.*, 2007; Nakazawa *et al.*, 2002; Tsien *et al.*, 1996b.

mice had deficits in a process called pattern completion (the retrieval of memories when only a fraction of the visual cues are presented) (Fig. 1.1G and H), and for memory acquisition of a one-time experience (Nakazawa et al., 2003). In line with these observations, Nakazawa and colleagues (2003) showed that place fields in these mice were significantly larger (e.g., less specific for a spatial location) in mutant mice as compared with control mice The DG is the third major region of the hippocampus that is part of the tri-synaptic circuit together with area CA1 and CA3. Making use of the proopiomelanocortin (PomC) promoter which is activated in the DG (Fig. 1.1A), the Tonegawa laboratory was able to investigate the role of DG NMDArs in perforant path LTP (Fig. 1.1I). McHugh et al. (2007) found that mice without NMDArs in the DG could not distinguish between two relatively similar contexts. Thus, they concluded that NMDArs in the DG are required to distinguish relatively similar environments, a process referred to as pattern separation (Fig. 1.1K). Together these studies showed that the same gene (NR1) was required for different aspects of memory formation in different subregions of the brain, an important conceptual advance made possible by the spatially restricted Cre/LoxP deletion system.

In addition to these studies focusing on the role of the NMDAr in memory storage, numerous other researchers used the Cre/LoxP system to study neurobiological processes. One of the ways that this system has been used is to ablate specific groups of neurons by crossing a Cre-expressing line with a second line that carries the loxP-STOP-loxP-IRES-diphtheria toxin fragment A (DTA). Cre expression mediates the excision of the floxed STOP codon leading to the expression of DTA resulting in cell death (Brockschnieder et al., 2004). Numerous mouse lines with floxed alleles as well as mutants that express Cre recombinase under control of a wide variety of promoters are becoming available (See for instance: http://www.mshri.on.ca/nagy/Cre-pub.html, http://www.jax. org, GENSAT http://www.gensat.org, and the Allen Brain Atlas http://www. brain-map.org). The combination of distinct Cre expression and floxed alleles allows researchers to generate unique mouse models to study the role of specific genes in a distinct subset of cells.

III. THE USE OF DOMINANT NEGATIVE INHIBITORS TO MANIPULATE GENE FUNCTION

As mentioned in the previous paragraphs, compensatory mechanisms hampered the interpretation of early mutant mouse models. Pharmacological inhibition of the activity of the cAMP-dependent PKA impaired LTP in hippocampal slices and suggested a role for PKA in memory storage (Frey et al., 1993; Huang and Kandel, 1994). However, analyses of knockout mice lacking the RIβ isoform of PKA showed no deficits in hippocampal LTP, but rather impairments in long-term depression (the weakening of neuronal connections) and a compensatory upregulation of the RI isoform (Brandon *et al.*, 1995). Thus, while the pharmacological approaches indicated an important role for PKA in memory storage, genetic studies failed to confirm the pharmacological observations.

A novel approach to determine the role of PKA in memory storage was taken by Abel *et al.* (1997). They generated a dominant negative form of the regulatory subunit of PKA referred to as R(AB). This was an inhibitor of both types of PKA catalytic subunits due to mutations in the two cAMP binding sites on the regulatory subunit (Clegg *et al.*, 1987; Ginty *et al.*, 1993). The advantage of using a dominant negative approach is that it is effective when multiple genes encode the same enzyme (as is the case for PKA) or when many homologues of the same gene exist (e.g., the coactivators CBP and P300, Vo and Goodman, 2001). Because the authors used the α-CaMKII promoter to drive expression of the transgene, the expression of the transgene was restricted to postnatal forebrain neurons (Fig. 1.2A) thereby avoiding developmental side effects or compensatory mechanisms. Using mice expressing R(AB) in forebrain neurons, Abel *et al.* (1997) examined the role of PKA in LTP in area CA1 of the hippocampus. They found that reduction of PKA activity and impaired LTP (Fig. 1.2B). The deficit in LTP was paralleled by behavioral deficits in spatial memory and in long-term, but not short-term memory for contextual fear conditioning (Fig. 1.2C), thus proving that PKA plays a critical role in the consolidation of long-term memory. In a later study, Rotenberg and colleagues (2000) showed that suppression of PKA activity in forebrain neurons impaired long-term place cell stability (Fig. 1.2D), an effect that could account for the memory deficits seen in these mice (Abel *et al.*, 1997).

IV. CONDITIONAL MANIPULATION OF GENE FUNCTION

A. The tTA system

Memory has several stages including acquisition, consolidation, and retrieval. These processes require the transient activation of specific cellular signaling pathways in particular brain regions. Previously formed memories can be modified further through reconsolidation and performance can change during extinction trials while the original memory remains intact (Abel and Lattal, 2001). To determine the role of specific genes in these different stages of memory, genetic-based systems were needed with a higher temporal resolution. To this end, the laboratory of Hermann Bujard developed a promoter based on the regulatory elements of the tetracycline resistance operon of *E. coli* (Furth *et al.*, 1994; Gossen *et al.*, 1994, 1995). In this operon, the tetracycline-controlled repressor (*tetR*) binds to its operator to repress the expression of resistance genes conferring

R (AB) expression in forebrain neurons

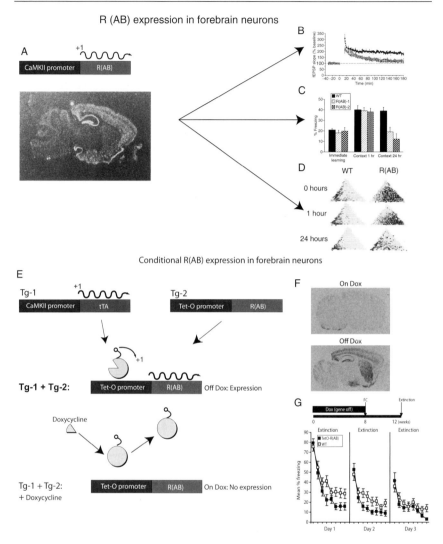

Figure 1.2. Genetic demonstration of a role of PKA in memory formation and LTP. (A) A sagital section of a transgenic mouse expressing R(AB) under control of the α-CaMKII promoter. The transgene is expressed throughout the hippocampus, cortex, olfactory bulb, amygdala, and striatum. (B) In two different lines of transgenic mice, L-LTP was reduced following four 100 Hz trains (1 s duration, 5 min apart) of stimulation. A stable level of robust potentiation was observed in wild-type controls (black circles), while a marked decline of potentiation back to near baseline values was measured in both lines of transgenic mice (open circles and squares). (C) R(AB) transgenics show a deficit in long-term, but not short-term memory for contextual fear conditioning. R(AB) mice and controls were given one CS/US pairing. No significant difference between mutants

survival in the presence of antibiotic. By fusing this promoter with the activation domain of virion protein 16 (VP16) of the herpes simplex virus *tetR* was converted to a transcriptional activator in eukaryotic cells. The product of this fusion protein, called the tetracycline-regulated transactivator (*tTA*), activates the transcription of transgenes containing the tetracycline operator (tetO). Importantly, the binding *tTA* to the tetO promoter was suppressed in presence of tetracycline or doxycycline (dox).

The laboratory of Eric Kandel used this newly developed *tTA* system to generate mice in which a transgene could be "conditionally" expressed in the mouse brain. The use of this system had the major advantage that transgene expression could be suppressed by administration of dox in the food or drinking water of animals. To study the role of α-CaMKII in learning and memory, they conditionally expressed a constitutive active form of α-CaMKII (Mayford *et al.*, 1996). They first developed a mutant mouse that expressed *tTA* under control of the α-CaMKII-promoter. This mutant was crossed to a second mutant in which the tetO promoter regulated expression of a mutant form of α-CaMKII that was calcium-independent, due to point mutation of a critical residue, Asp286. In bi-transgenic offspring, expression of the constitutive active form of CaMKII resulted in the loss of hippocampal LTP induced by 10 Hz stimulation, as well as the loss of spatial memory. Importantly, delivery of dox resulted in the suppression of transgene expression, leading to the restoration of normal LTP and

and controls was observed either immediately or 1 h after the CS/US pairing. However, 24 h after the training session, freezing responses of both R(AB) transgenic lines were significantly reduced compared with wild-type mice. (D) Examples of place fields in wild type and R(AB) transgenic mice. In contrast to wild types, R(AB) transgenic mice have unstable place cells. (E) Schematic representation bitransgenic systems to induce R(AB). Two different transgenes are used: one expressing the tetracycline-responsive transactivator (tTA) under the control of the αCaMKII promoter (Mayford *et al.*, 1996) and another carrying the R(AB) cDNA under the control of tet operator sequences (tetO promoter). In mice bearing both transgenes, expression of R(AB) is suppressed when mice are fed with dox-containing chow. (F) Under doxycycline conditions, bitransgenic mice do not express R(AB). Bitransgenic mice raised on doxycycline and removed from doxycycline for 28 days showed a high level of R(AB) mRNA, as detected by *in situ* hybridization in hippocampus, cortex, striatum, and amygdala. (G) Under these conditions, transgenic reduction of PKA activity in forebrain neurons in the adult facilitates the extinction of contextual fear. Upper panel, both wild type and bitransgenic mice were raised on doxycycline to suppress R(AB) expression, fear conditioned, and removed from doxycycline 24 h later to permit R(AB) expression in the adulthood. After 28 days, mice were subjected to contextual extinction. Lower panel, R(AB) expressing mice and wild-type littermates were reexposed to the conditioning context in 24 min sessions for three consecutive days. Freezing is shown in blocks of 3 min for each 24 min daily session. Error bars indicate SEM. *Sources:* Abel *et al.*, 1997; Isiegas *et al.*, 2008.

memory (Mayford *et al.*, 1996). Thus, this pivotal study showed that the loss of LTP and spatial memory was due to adult-specific expression of a transgene and not to developmental side effects. In the years thereafter, the tTA system has been used by many other laboratories to conditionally manipulate molecular signaling pathways in specific tissues or cell types.

Conditional genetic systems opened a new era of research that was not possible using constitutive transgenic or knockout approaches. Investigators could now establish the role of specific transgenes in memory retrieval, reconsolidation as well as memory extinction. For instance, the Abel laboratory used this conditional strategy to determine the role of PKA in the extinction of previously acquired contextual fear memories. For this purpose, they generated bi-transgenic mice in which R(AB) expression was controlled by the tTA system (Fig. 1.2E), thus enabling them to suppress R(AB) expression under dox conditions (Fig. 1.2F). During development and continuing through training in the contextual fear conditioning paradigm, mice were fed dox to suppress the expression of the R(AB) transgene. After training, the animals were removed from dox, initiating the expression of the PKA inhibitor. After 4 weeks, the rate of extinction of the previously acquired contextual fear memory was assessed. The researchers found that reducing PKA activity selectively during the extinction phase slowed extinction, suggesting that PKA functions as a constraint on the erasure (erasure or formation of extinction memory) of previously acquired fear memories (Isiegas *et al.*, 2006). Using the same conditional approach, a later study by Havekes *et al.* (2008) demonstrated that conditional suppression of forebrain calcineurin (also known as protein phosphatase 2B) facilitated the extinction of contextual fear memories. This was of particular interest because PKA and calcineurin target similar substrates and function antagonistically. Thus, the conditional tTA system has allowed analysis of gene function during distinct types of memory formation.

A second major use of this conditional system has been to manipulate gene function selectively during development, while leaving it undisturbed during adulthood (by putting the mice on dox starting from 2 months after birth). In a recent paper by Kelly and colleagues (2008), overexpression of the G-protein subunit Gαs, a protein subunit that stimulates adenylyl cyclase activity is genetically linked to schizophrenia (Avissar *et al.*, 2001; Memo *et al.*, 1983; Minoretti *et al.*, 2006), selectively during development induced several schizophrenia-related phenotypes. They found impairments in spatial learning and memory formation in adult mice using the Morris water maze when the transgene was only expressed during development. In addition, they found that the ventricles of the mice were enlarged, while the ventral and dorsal striatum size were reduced as a consequence of overexpression of the Gαs subunit selectively during development. These experiments, along with others that examined anxiety-related behaviors (Gross *et al.*, 2002), demonstrate that certain endophenotypes of psychiatric disorders are likely the result of alterations in the function of neuronal circuits that occur during development.

In addition to the use of the tTA system to study the role of specific genes in neurons, the laboratory of Philip Haydon has used this system to determine the role of astrocytes in synaptic plasticity, modulation of sleep homeostasis, and the cognitive consequences of sleep loss. The laboratory first generated a transgenic mouse line in which tTA was driven by the astrocyte-specific GFAP to obtain astrocyte-specific expression of tTA (Pascual *et al.*, 2005). Next, they generated a transgenic mouse that expressed the cytosolic portion of the N-ethylmaleimide-sensitive factor attachment protein (SNARE) domain of synaptobrevin 2, under control of the tetO promoter. A previous study by the same laboratory showed that overexpression of this portion of the SNARE protein blocks gliotransmission, another example of using a dominant negative approach (Zhang *et al.*, 2004). By breeding the GFAP-tTA line and tetO-SNARE line they were able to produce double transgenic mice with vesicular transmission blocked selectively in astrocytes. They found that gliotransmission critically regulated synaptic plasticity and had a previously unsuspected role in sleep regulation. Inhibition of gliotransmission attenuated the accumulation of sleep pressure, the drive to go to sleep, and prevented cognitive deficits associated with sleep loss, thus providing evidence that astrocyte signaling regulates the relationship between sleep and memory formation (Halassa *et al.*, 2009).

B. A novel form of the tTA system: The rtTA system

Although the tTA system has great advantages in comparison with constitutive transgenic systems, this system requires that mice be kept on dox during breeding and at least the first weeks after birth to suppress transgene expression during development. To avoid this long period of dox exposure which could lead to developmental side effects and create an imbalance of the intestinal flora resulting in diarrhea and in a smaller number of animals, a reversed version of tTA was developed by the Kandel group called rtTA. In case of rtTA, dox treatment results in activation of a tetO-driven transgene, rather than its suppression (1998). Using this novel system, mice had to be kept on dox for a minimum of 1 week prior to the start and during the course of an experiment. Using this modified form of tTA, Mansuy and colleagues (1998) showed that forebrain-specific overexpression of the catalytic subunit of calcineurin, the calcium-dependent protein phosphatase calcineurin, impairs the formation of spatial memories and an intermediate form of LTP. In agreement with this finding, Malleret and colleagues (2001) found that conditional overexpression of an inhibitor of calcineurin facilitates memory formation and enhanced LTP. This phenotype was reversed when animals were taken of dox. In a more recent study, the laboratory of Isabelle Mansuy elucidated that conditional inhibition of a downstream target of calcineurin, the protein phosphatase 1 (PP1) enhances the formation of spatial memories using various learning tasks as well as LTP and that this enhancement could be reversed by taking the animals of dox (Genoux *et al.*, 2002).

V. PHARMACOGENETIC APPROACHES

A. Tamoxifen-controlled gene manipulation

The application of the Cre/loxP system in combination with promoters that are expressed selectively during adulthood allowed researchers to knockout genes during adulthood avoiding developmental side effects or even during specific stages of memory, but the temporal resolution was still relatively poor. To improve this temporal resolution, Cre was fused to the mutated ligand-binding domain of the human estrogen receptor (ER) (Feil *et al.*, 1997; Metzger *et al.*, 1995a,b). As a consequence of the mutation, the receptor does not bind endogenous estradiol, but is highly sensitive to tamoxifen. Under conditions without tamoxifen, the fusion product (referred to as CreERT2) is trapped in the cytoplasm by binding to heat shock proteins (Brocard *et al.*, 1997). By injections of tamoxifen, the fusion product is released from the complex and translocates to the nucleus and catalyzes the recombination of loxP sites resulting in deletion of the flanked gene (Erdmann *et al.*, 2007; Metzger *et al.*, 1995b). The use of CreERT2 has two substantial advantages: (1) gene manipulation can be restricted to distinct populations of cells via the local delivery of tamoxifen and (2) application of tamoxifen causes Cre-mediated gene excision on the order of days (Brocard *et al.*, 1997). Imayoshi *et al.* (2008) used this system to study the role of continuous neurogenesis in the structural and functional integrity of the adult forebrain. They generated transgenic mice in which the nestin promoter was used to drive CreERT2 expression in neural stemcells (Imayoshi *et al.*, 2006) and crossed this transgenic line with a second transgenic line in which carried the loxP-STOP-loxP-IRES-DTA, driven by the neuron-specific *enolase2* gene. Importantly, tamoxifen treatment in bi-transgenic adult mice led to the deletion of the STOP codon in DTA cassette in neuronal stem cells. As a result, DTA was expressed as soon as cells started to differentiate into neurons, thereby killing them. This resulted in impaired neurogenesis and the specific loss of granule cells in the olfactory bulb and DG of the hippocampus. Thus, rather than deleting a gene, in this study the Cre/loxP system was used to cause expression of a gene as opposed to knocking one out. Behavioral screening of transgenic mice that received tamoxifen indicated that loss of these granule cells resulted in impairments in spatial memories (as indicated by deficits in the retention test of the spatial version of the Morris water maze and Barnes maze) and in contextual memories (Imayoshi *et al.*, 2008). These studies by Imayoshi *et al.* (2008) demonstrated that adult neurogenesis is required for the modulation and refinement of the existing circuits in the DG important for hippocampus-dependent memory formation.

The tamoxifen-inducible regulation conferred by fusion of the ERT domain onto a protein of interest has also been applied to tightly control the time of transcription factor activation during memory formation. To temporally manipulate functioning of the cAMP-response element-binding protein (CREB, a transcription factor), Kida and co-workers (2002) fused an α-CREB isoform with a mutation in the S133 site (a form of CREB that cannot be activated by phosphorylation of the S133 site) to a ligand-binding domain of the mutant human ER. In the absence of tamoxifen, the fusion protein is inactive (Feil *et al.*, 1996). However, administration of tamoxifen activates the inducible CREB-repressor fusion protein (CREBIR), allowing it to compete with endogenous CREB and disrupt cAMP-responsive element (CRE)-mediated transcription (Kida *et al.*, 2002). Using this inducible and reversible CREB repressor system, they showed that CREB is essential for the consolidation of long-term conditioned fear memories. Administration of tamoxifen 6 or 12 h, but not 24 h or 30 min before fear conditioning training impaired memory formation indicated by reduced freezing levels during the 24 h retention test. Importantly, the finding that CREBIR mice administered tamoxifen more than 12 h prior to fear conditioning showed no deficit in subsequent freezing during the 24 h retention test demonstrated that the impairment produced by tamoxifen in CREBIR mice is behaviorally reversible.

Using a similar approach, Li *et al.* (2007) generated an inducible and reversible mutant mouse model to study the role of the "Disrupted-in-schizophrenia 1" (DISC1) gene, which is thought of playing a pivotal role in the development of schizophrenia pathogenesis. To do this, they generated mice expressing DISC1-cc, a dominant negative form of the DISC1 protein, under control of the α-CaMKII promoter, which lead to expression of the transgene selectively in forebrain neurons. The DISC1-cc protein was fused to the mutant ER, which can be activated by tamoxifen. Without the presence of tamoxifen, the DISC1-cc transgene is sequestered into inactive complexes by heat shock chaperone proteins. When tamoxifen is present, DISC1-cc is released from the ligand-binding domain and competes with wild-type DISC1 protein for the binding of Nudel and Lis1, genes that regulate several aspects of brain development. However, the effect of tamoxifen is transient: as soon as tamoxifen is metabolized DISC1 signaling returns to normal. This inducible and reversible transgenic system enabled Li and co-workers to study the impact of disruption of DISC1 function on specific time points during brain development. Their study shows that induction of a mutant DISC1 protein early in postnatal development is sufficient to impair spatial working memory and other depressive like traits that parallel changes associated with schizophrenia-related DISC1 sequence variations in humans. Thus, with these studies, Kida *et al.* (2002) and Li *et al.* (2007) demonstrated the power of using tamoxifen-inducible transgenic systems to study complex cognitive processes.

B. Pharmacogenetic regulation of neuronal excitability

In recent years, new technologies have been developed to control the activity of genetically restricted neural populations with high temporal resolution. One of the pharmacogenetic approaches used to suppress neuronal activity, through the enhancement of chloride currents, is the expression of the Ivermectin-gated chloride channels from C. *elegans*. Lerchner and colleagues (2007) first coexpressed the two subunits (GluClα and GluClβ) that together form the Ivermectin-gated chloride channel unilaterally in the striatum of rats using viral injection. Next, they tested whether unilateral activation of the chloride channels in the striatum using Ivermectin treatment perturbs of rotational behavior induced by amphetamine. As expected, mice in which the chloride channels were activated in one hemisphere showed a marked increase in the net number of rotations towards the injection side as a consequence of the imbalance of neuronal activity between the striatal regions in the two hemispheres. Wild-type animals did not show any preference for the direction of the rotations. After 4 days, the effect of Ivermectin was fully reversed. Thus, with this study the authors show that Ivermectin-gated chloride channels can be used to temporally inactivate a restricted population of cells over a relatively short time period.

An alternative approach to inhibit neuronal activity was developed by Tan *et al.* (2006). They used the *Drosophila* allatostatin receptor to reduce neuronal activity in a specific population of cells. Binding of allatostatin to its receptor results in the opening of the G-protein coupled inward rectifier potassium channel (GIRK) leading to hyperpolarization of the cell. Gosgnach *et al.* (2006) used the allatostatin receptor to temporally block the activity of V1 motor neurons to define the function of these neurons in motor behavior. They found that V1 motor neurons regulate the speed of vertebrate locomotor movements.

A third approach to inhibit neuronal function is via manipulation of GABAA receptor activity. GABAA receptors can be positively modulated by the drug zoldipem. Application of this drug results in the facilitation of GABA receptor-mediated transmission, inhibiting excitatory responses (thereby reducing neuronal activity). Wulff and colleagues (2007) manipulated the sensitivity of the GABAA receptor by first generating a line of mice that expressed a mutated gamma2 subunit of the GABAA receptor that is insensitive to zoldipem. Next, they reintroduced the wild-type subunit in a genetically restricted group of neurons by using Cre recombinase to swap the wild-type subunit in place of the mutant allele to restore zoldipem sensitivity. Bath application of zoldipem in cerebellar slices of mice expressing the wild-type receptor selectively in Purkinje cells resulted in the facilitation of inhibitory postsynaptic currents in cerebellar Purkinje cells. Intraperitonal injections of zoldipem in these mutant mice resulted in impairments in motor coordination as measured using the rotarod task (Wulff *et al.*, 2007).

An alternative approach to the methods described above is the application of molecules for inactivation of synaptic transmission (MISTs). MISTs are modified presynaptic proteins that interfere with the synaptic vesicle cycle when cross-linked by small molecule "dimerizers" (Karpova *et al.*, 2005). Karpova and colleagues (2005) showed that MISTs based on VAMP2/Synaptobrevin and Synaptophysin could rapidly (e.g., within 10 min) and reversibly inhibit synaptic transmission. As a proof of principle, they expressed the MISTs selectively in Purkinje neurons. Mice were acquainted with the rotarod task in the absence of dimerizer. Next, they injected dimerizer in the lateral ventricle during or following the acquisition of the task. Rotarod performance was significantly impaired in the presence of the dimerizer. The effect was fully reversed 36 h after the injection. Thus, with this study, they show that MISTs can be successfully used to specifically perturb the function of specific neuronal circuits *in vivo* with temporal precision in the order of hours.

In summary, numerous new sophisticated methods are emerging that permit temporary reduction in the activity of a genetically restricted group of cells and demonstrate that acute silencing of selected populations of neurons can be used to elucidate their function in well-defined behaviors. However, these techniques cannot be used to elucidate the function of specific cellular signaling pathways in learning, memory formation, and its retrieval, since these methods regulate neuronal activity rather than manipulating a specific intracellular signaling pathway. Thus, specific pharmacogenetic approaches are needed that rapidly modulate specific intracellular signaling processes within subsets of neurons.

C. Targeting of intracellular signaling pathways using pharmacogenetic approaches

Biochemical processes within neurons are activated with a time course of minutes to hours during memory storage. Further, distinct biochemical processes underlie each of the different stages of memory including acquisition, consolidation, and retrieval (Abel and Lattal, 2001). Many studies have tried to unravel the temporal dynamics of distinct cellular signaling pathways using pharmacological approaches. Although these approaches allow for fast manipulation of cellular signaling pathways, they are not specific for distinct cell types.

To circumvent this problem, novel tools have been generated that combine the cell type and regional specificity possible with transgenic techniques together with the high temporal resolution needed to activate specific molecular pathways in distinct groups of cells.

The Abel laboratory developed such a tool to transiently manipulate the cAMP pathway specifically in neurons by taking advantage of heterologous G-protein-coupled receptors found in invertebrates (Isiegas et al., 2008). They expressed a Gαs-coupled *Aplysia* octopamine receptor (Ap oa1) under control of the tTA system. Activation of the receptor leads to the activation of the Gs-signaling pathway, resulting in a rapid and transient increase in levels of the intracellular second messenger cAMP specifically in forebrain neurons due to the use of α-CaMKII promoter to drive tTA expression (Fig. 1.3A). The use of this Ap oa1 had two major advantages. (1) Octopamine is present only at trace levels in the mammalian central nervous system; thus, the Ap oa1 is only active when exogenous octopamine is supplied. (2) Octopamine does not act on other receptors in the mammalian central nervous system; thus, expression of the Ap oa1 is required for an effect of exogenous octopamine on Gαs signaling. LTP is facilitated in hippocampal slices from Ap oa1 mice after treatment with octopamine (Fig. 1.3B), and transient activation of the Ap oa1 *in vivo* facilitates the formation of long-term memories for contextual fear (Fig. 1.3C), and also facilitated the retrieval of previously formed memories (Fig. 1.3D). Overall, this study demonstrated the usefulness of using the octopamine receptors to manipulate specific intracellular pathways in a conditional and spatially restricted manner.

Besides the use of heterologous octopamine receptors to examine the role of distinct second messenger pathways in the mammalian system, other genetically engineered G-protein coupled receptors have been developed to combine genetic with pharmacological approaches. For example, Sweger and colleagues (2007) developed a transgenic mouse that expresses a genetically engineered receptor that is activated solely by a synthetic ligand (receptor activated solely by a synthetic ligand, RASSL). The ro1 receptor is a κ-opiod receptor (KOR) modified by replacing its second extracellular loop with the second extracellular loop of the δ-opiod receptor, such that receptor activation causes reduced cAMP-levels via the Gi-pathway. As a consequence of this modification, the affinity of this receptor for its endogenous ligands is reduced, while it maintains the ability to bind the synthetic ligand spiralodine. By conditionally expressing this receptor in astrocytes using the tTA system cells on a KOR knockout background, the authors were able to manipulate Gi-signaling specifically in glial cells while measuring changes in neuronal excitability. They found that expression of the modified KOR selectively in astrocytes induces the development of hydrocephalus (the abnormal accumulation of cerebrospinal fluid in the brain). This study shows the potency of

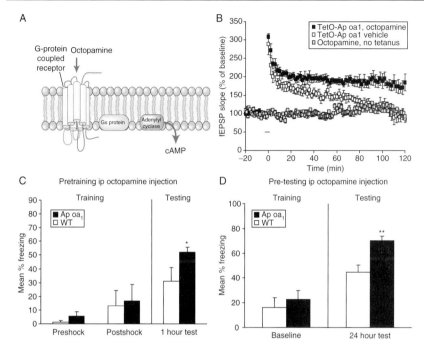

Figure 1.3. Conditional activation of the *Aplysia* octopamine receptor in forebrain neurons facilitates memory formation and LTP. (A) Activation of the *Aplysia* octopamine receptor by its natural ligand activates the Galphas signaling pathway to transiently increase cAMP. (B) Treatment of hippocampal slices from Ap oa1 mice with octopamine induces long-lasting potentiation following stimulation that normally elicits transient potentiation. (C) Training and 24 h long-term memory test during contextual fear conditioning in Ap oa1 mice and wild-type littermates injected intraperitoneally with octopamine 30 min before training. No differences in freezing behavior are observed between groups before and directly after the shock during training. In contrast, Ap oa1 mice show a significant increase in freezing behavior when reexposed to the fear-conditioned context 24 h after training. (D) Intraperitonal injections of octopamine 30 min prior to the retention test facilitates memory retrieval in Ap oa1 mice expressing the *Aplysia* octopamine receptor. *Source:* Isiegas *et al.*, 2008.

using RASSLs that can be activated solely by synthetic drugs. Indeed, many new RASSLs have recently been developed that target specific second messenger systems *in vivo* (Conklin *et al.*, 2008), allowing for the manipulation of other G protein-coupled receptor pathways in addition to the Gi-signaling pathway.

The Tsien laboratory recently examined the role of α-CaMKII via pharmacogenetic manipulation (Cao *et al.*, 2008; Wang *et al.*, 2008). First, the authors generated mice that expressed a constitutively active form of α-CaMKII in forebrain neurons (using the α-CaMKII promoter). As a result, the overall activity of α-CaMKII is elevated in forebrain neurons, but the activity of this mutant form of α-CaMKII is inhibited by a small inhibitor molecule called NM-PP1 that does not affect any other proteins in the brain. Using this system, the Tsien laboratory could manipulate αCaMKII activity within the range of minutes by simply injecting mice with NM-PP1 allowing them to explore the role of α-CaMKII activity levels during distinct phases of memory and LTP. They found that the initial 10 min of memory formation and LTP are sensitive to inducible genetic downregulation of α-CaMKII activity, suggesting that molecular dynamics of α-CaMKII play an important role in the representation of short-term memory during this critical time window (Wang et al., 2008).

VI. OPTOGENETIC APPROACHES

Great progress has been made in understanding how neural network activity underlies the formation of memories. Many of the previously discussed approaches allow the study of involvement of activity or signaling cascades on a timescale of minutes to hours. However, neuronal and network activity works on a millisecond timescale. Thus, several research laboratories started to search for ways to manipulate neuronal activity with a time-scale of milliseconds. One successful approach used to manipulate the activity of groups of cells with the desired time-scale is the use of caged neurotransmitters that can be released upon stimulation with light (a process called photo-uncaging). For example, Tanaka et al. (2008) demonstrated the necessity of protein synthesis and brain derived neurotrophic factor for spine enlargement of CA1 neurons, a process thought to underlie the strengthening of synapses, by combining two-photon photo stimulation to locally uncage glutamate in combination with synaptic stimulation. Harvey and Svoboda used two-photon uncaging of glutamate and synaptic stimulation to show that after LTP-induction in a specific spine, weak stimulation can induce strong LTP in neighboring spines (Harvey and Svoboda, 2007). The disadvantage of this approach is that it lacks cell type specificity and cannot be used in vivo to stimulate cells in deeper layers of the brain.

　　The discovery of two light-sensitive rhodopsin channels in the unicellular green alga Chlamydomonas reinhardtii (Nagel et al., 2002, 2003; Sineshchekov et al., 2002) allowed for novel ways to manipulate neural function within a time-scale of milliseconds. Channelrhodopsin-1 is a proton-gated (H^+) channel (Nagel et al., 2002), whereas Channelrhodopsin-2 (ChR2) is a light-gated cation channel (Nagel et al., 2003). Initial in vitro studies demonstrated the feasibility of using ChR2 to manipulate neural activity using light. Using lentiviral gene delivery to express ChR2 in mammalian neurons, Boyden et al. (2005) showed that with a series of brief pulses of light, ChR2 could reliably mediate trains of spikes or synaptic events including excitatory and inhibitory synaptic transmission with millisecond-timescale temporal resolution. Adamantidis and colleagues (2007) applied this optogenetic tool *in vivo* to manipulate the activity of hypocretin-expressing neurons in the lateral hypothalamus (de Lecea *et al.*, 1998; Peyron *et al.*, 1998) to examine the causal relation between activity of this distinct group of neurons and arousal stability. They specifically targeted hypocretin-neurons, using a lentivirus that drives the expression of a ChR2 construct under control of the hypocretin promoter (Fig. 1.4A), in mice chronically implanted with electroencephalographic (EEG) and electromyographic (EMG) electrodes (Fig. 1.4B) to measure brain and muscle activity, respectively, to monitor the sleep/wake state of mice. A 200-M optical fiber was used to deliver laser light into the lateral hypothalamus of freely moving mice. They first

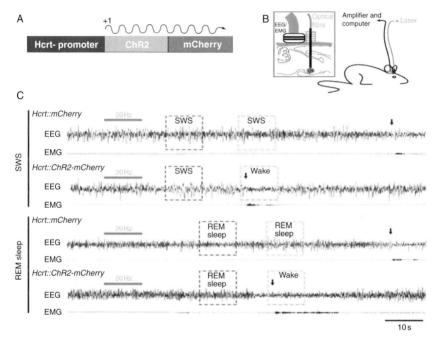

Figure 1.4. Optogenetic manipulation of the neural substrates of awaking. (A) A diagram of the construct used to drive ChR2 specifically in hypocretin-producing neurons. (B) Schematic of the behavioral setup used for *in vivo* deep-brain photostimulation in mice. Magnification (inset) shows the EEG/EMG connector used for sleep recording and the cannula guide used for lateral hypothalamus light delivery through an optical fiber. (C) Representative EEG/EMG recordings showing reduced time to awakening after a single bout of photostimulation (15 ms, 20 Hz, 10 s) of hypocretin neurons in mice expressing the ChR2. Light stimulations are represented by horizontal blue bars. Awakening events are indicated by vertical black arrows. *Source:* Adamantidis *et al.*, 2007. (See Page 1 in Color Plate Section at the end of the book.)

investigated whether light stimulation facilitated protein expression of the immediate early gene FOS. They found that 10 s of 20 Hz delivered once per minute over 10 min increased the number of Fos expressing Hcrt-neurons from 25 to 65%, elegantly demonstrating that stimulation of ChR2 channels with light *in vivo* facilitates neuronal activity. Next, Adamantidis *et al.* (2007) stimulated the hypocretin-neurons that expressed the ChR2 channels with light during distinct phases of sleep (as defined by the EEG recordings). They determined the latency between the end of the photo-stimulation and the next transition to wakefulness. They found that 10 s of light stimulation with a frequency of 5–20 Hz or continuous light markedly reduced the latency to waking up from slow wave sleep (Fig. 1.4C). Similarly, photo-stimulation with a frequency of 5–30 Hz shortened the latency to waking from rapid eye movement sleep (a sleep state that is characterized by high frequency activity waves in the brain) in ChR2 expressing mice (Fig. 1.4C). Thus, using this sophisticated optogenetic approach, Adamantis *et al.* (2007) were able to establish a causal relationship between frequency-dependent activity of a genetically defined neural cell type and specific mammalian behavior central to clinical conditions and neurobehavioral physiology. This pioneering study paved the road for future studies to determine how specific activity patterns of a genetically defined group of neural cells underlie the formation and storage of long-term memories.

VII. VISUALIZATION OF BRAIN STRUCTURE AND FUNCTION

During the twentieth century numerous new molecular techniques and approaches became available to study the role of specific genes in behavior *in vivo*. The biochemical and genetic approaches developed were groundbreaking and pushed the neuroscience field rapidly forward. Yet, there was a lack of tools that allowed for the visualization of intra- and extra-cellular processes, including protein–protein interactions. A major step towards the development of such a tool was made by Osamu Shimomura, Martin Chalfie, and Roger Tsien by the discovery and development of the green fluorescent protein (GFP) for which they were awarded the Nobel Prize in 2008. The GFP gene was cloned and described in 1992 (Prasher *et al.*, 1992). Many researchers attempted to heterologously express GFP in various model systems without success. Martin Chalfie discovered that the original cDNA isolated by Prasher contained an inhibitory sequence preventing GFP from being expressed. By deleting the inhibitory sequence, Chalfie succeeded in expressing GFP in various model organisms including *C. elegans* leading to a publication in science in 1994 (Chalfie *et al.*, 1994). In the years thereafter, many new forms GFP were developed including various spectral variants (Heim and Tsien, 1996; Heim *et al.*, 1994, 1995;

Zacharias and Tsien, 2006). In the paragraphs below, we will describe some of the most recently developed tools to visualize the processes underlying memory storage and synaptic plasticity.

A. The TetTag system

A very popular technique to determine these activity profiles of neuronal populations is to examine the expression of immediate early genes like *c-fos* using immunohistochemistry. Immunohistochemistry was first described in 1941 by Coons *et al.* (1941) (for review see LeDoux, 2000; Ramos-Vara, 2005) and relies on the binding of specific antibodies to antigens of interest. The antigen–antibody binding is visualized by histochemical reactions or fluorochromes. The development of phospho-specific antibodies that bind specifically to the activated form of a protein increased the potency of this technique. One of the most frequently used phospho-sites to map neuronal activity is the serine 133 residue of the CREB, a transcription factor that is crucial for memory consolidation. Upon phosphorylation of this site, the transcription factor CREB initiates gene transcription of CREB target genes (Mayr and Montminy, 2001). By monitoring CREB phosphorylation using immunohistochemistry, researchers have been able to monitor when and where CREB activity occurs during after learning or memory retrieval. The disadvantage of this technique is that you can only assay the activity patterns at one particular time point, making it impossible to determine the activity profiles of the same cells at different times during learning, memory formation, and its retrieval. Alternative approaches including fMRI and small animal PET currently do not yet have the desired resolution to look at individual neurons. To circumvent these problems, a novel transgenic system (called the TetTag system) was developed recently by the Mayford laboratory that allows persistent tagging of neurons activated during a time window enabling them to determine whether this subset of neurons that were activated during learning were reactivated during retrieval (Reijmers *et al.*, 2007). This system is based on the combination of two novel transgenes. First, they generated a transgenic mouse expressing tTA under control of the promoter of the immediate early gene *c-fos*. Using this transgenic line, tTA is transiently expressed in those cells in which the *c-fos* promoter is activated (Fig. 1.5A). The second part of the TetTag system consists of a transgenic mouse containing a LacZ reporter and a dox-insensitive form of tTA (tTA*) both under control of a bi-directional tetO promoter. Double transgenic mice, referred to as TetTag mice, are raised on dox. As a consequence of dox treatment, neurons do not express tTA* and tau-LacZ (Fig. 1.5A, left yellow block). The time window for tagging is opened by taking the mice of dox which will initiate the activation of the tetO promoter in those neurons in which the *c-fos* promoter is activated, resulting in the expression of tTA* and tau-LacZ in this specific population of neurons (Fig. 1.5A,

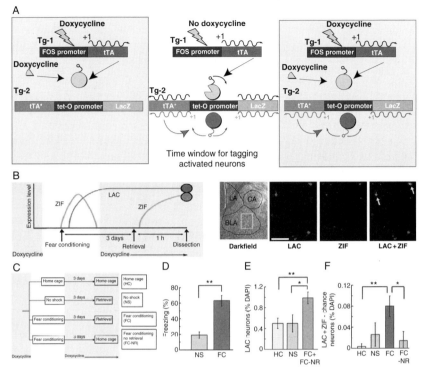

Figure 1.5. Tagging and visualization of previously activated neurons using the TetTag mouse. (A) TetTag mice are raised on food containing Dox (left yellow block). During this time, neuronal activation that leads to expression of tTA through Fos promoter activation will not trigger tagging, because doxycycline blocks activation of the tetO promoter. The time window for tagging is opened by switching mice to food without doxycycline (middle white block). Neuronal activation will now activate the transcriptional feedback loop and start expression of tau-LacZ. The time window is closed by putting mice back on doxycycline food (right yellow block) to block further feedback loop activation. However, neurons that were activated during the "no doxycycline" time window will continue to express tau-LacZ, because the feedback loop can maintain its own activation through the doxycycline-insensitive form of tTA. (B, left panel) A protocol was designed to detect repeated activation of neurons during learning and retrieval of conditioned fear. Learning takes place in the absence of doxycycline, when activation of neurons triggers long-lasting expression of tau-LacZ (LAC) and short-lasting expression of immediate-early genes like Zif/Egr (ZIF). Retrieval takes place 3 days after learning in the presence of doxycycline, preventing activation of the molecular feedback loop. Dissection of brains is done 1 h after retrieval, when LAC and ZIF can be used as indicators of learning-induced and retrieval-induced activation, respectively. (B, right panel) Example of LAC and ZIF expression in BLA neurons of a mouse that was subjected to the protocol described in (A). The yellow square in the left dark field picture marks the area shown in the three immunostaining pictures. Yellow arrows mark neurons that express both LAC and ZIF. (C) A diagram of the experimental design.

white panel). This initial expression of tTA* generates a transcriptional feedback loop that maintains its own activity. By putting the mice back on dox further tagging of neurons is prevented (Fig. 1.5A, right yellow panel). Using this system, Reijmers *et al.* were able to show that neurons activated and tagged during learning are reactivated during retrieval suggesting that these neurons may encode memory itself (Fig. 1.5B–F). The TetTag system will allow this possibility to be further tested by selective expression of tetO-regulated transgenes in these specific neuronal populations after learning.

In a second study, the Mayford laboratory wanted to answer the question how new proteins, synthesized in the soma exert their effect on specific synapses involved in synaptic or behavioral plasticity (Matsuo *et al.*, 2008). To test the hypothesis that newly synthesized AMPA receptors are recruited to specific synapses, the Mayford laboratory first generated a mouse that expresses a GluR1 subunit, the main subunit of the AMPA receptor, fused to GFP under control of the tetO promoter and crossed this mouse line with the *c-fos*-tTA mouse line. This combination, referred to as GFP-GluR1$^{c\text{-}fos}$, allowed them to express the GluR1-GFP subunit in a dox regulated and neuronal activity dependent manner. Using this system, Matsuo *et al.* (2008) showed that learning induces the synthesis of GluR1 subunits that are recruited selectively to the mushroom shaped spines.

B. The Brainbow mouse

Santiago Ramón y Cajal pioneered the study of neurobiology by describing the neuroanatomical structure of neurons and their connections using Golgi's silver stain to label small numbers of neurons. Although this technique could be used to examine the connections of small groups of neurons, its usage was limited in the case of high density populations. To overcome this problem, a multicolor Golgi technique needed to be established that would allow researchers to map many neurons within a single brain section.

Livet *et al.* (2007) developed a new approach taking advantage of the Cre/LoxP system (Branda and Dymecki, 2004 see also Section II.A), that can alter gene expression by DNA insertion, deletion, or recombination. They first generated several constructs containing the DNA of different variants of fluorescent proteins which were available, including yellow fluorescent protein (YFP), cyan fluorescent protein (CFP), red fluorescent protein (RFP), and orange

---

(D) The FC group showed more freezing than the NS group during context retrieval. (E) The combined FC and FC-NR group had an increased number of LAC-positive neurons as compared with both the HC and NS groups. (F) The FC group had more LAC + ZIF neurons than the HC and FC-NR groups. *Source*: Reijmers *et al.*, 2007. (See Page 1 in Color Plate Section at the end of the book.)

fluorescent protein (OFP) (Shaner *et al.*, 2004). In these constructs, referred to as Brainbow 1.0 and 2.1, tamoxifen-induced Cre recombinase results in the excision and/or inversion of tandem DNA segments encoding the different fluorescent proteins due to the presence of incompatible loxP variants. Cells expressing the *Brainbow-2.1* construct in which Cre expression could mediate three different inversions and two excisions showed stochastic expression of all different forms of fluorescent protein upon the delivery of Cre (e.g., OFP, YFP, MFP, and RFP).

Next, they generated transgenic mouse lines in which the different Brainbow constructs were expressed under control of the regulatory elements of the *Thy1* gene, a promoter that drives high levels of expression in a variety of neurons (Caroni, 1997; Feng *et al.*, 2000). The founder mouse was mated with other transgenic lines expressing the tamoxifen-inducible Cre. Cre-mediated excision and/or inversion was achieved by giving pups a single injection of tamoxifen on the day of birth (P0). Inversion and excision recombination events led to several distinct colors due to stochastic recombination. Besides the traditional fluorescent colors, they also found the coexpression of multiple colors in individual cells. The presence of polychromatic cells was a consequence of the method used for the generation of the transgenic lines. Pronuclear injection of constructs can lead to tandem integration of multiple transgene copies (Palmiter and Brinster, 1986) and Cre mediates recombination in an independent manner in these distinct copies. This led to a mosaic color palette with up to 89 distinguishable colors in a transgenic mouse line expressing the *Brainbow-1.0* construct. The distinct color patterns of adjacent neurons allowed for visualization of cellular interactions on a large scale in regions of high neuronal density. Currently, new methods to section, image, and analyze entire mouse brains are being developed by the Jeff Lichtman and others in hopes of generating a complete map of all connections within the mouse brain.

VIII. THE CHALLENGES FACING GENETIC APPROACHES IN THE MOUSE

Over the past decade, many novel tools and methods have been generated to alter gene function *in vivo*. Although these techniques have been greatly refined to allow researchers to manipulate gene function in a spatially and temporally restricted manner, and thus avoid behavioral side effects unrelated to learning, caveats still exist that need to be considered when interpreting results using these techniques.

Leakiness of a transgene can occur due to loss of the specificity of the promoter depending on the site of integration resulting in a lack of temporal control over transgene expression. Furthermore, disruption of an unknown gene as a consequence of transgene insertion can lead to a phenotype that is partly or entirely due to loss of the unknown gene, rather than expression of the transgene itself. For that reason, multiple transgenic lines carrying the same construct should be tested. If the different mutants have the same phenotype, one can assume that the phenotype is a consequence of the gene manipulation and not the loss of an unknown gene. In case of the tTA system, this issue could be addressed by looking at the phenotype of mice in the presence of dox. Under dox conditions, transgene expression should be suppressed, thus the phenotype should be lost. If the phenotype is still present under dox conditions then this would argue that the phenotype is a consequence of the loss of an unknown gene due to the location where the transgene is inserted.

Certain promoters are more broadly expressed during development, but in a more restricted fashion during adulthood. Some studies have used these promoters to manipulate genes selectively in the regions in which the promoter is expressed during adulthood. For example, the KA1 promoter can be used to manipulate gene function specifically in hippocampus subregions (Nakashiba *et al.*, 2008; Nakazawa *et al.*, 2002, 2003), although the promoter is also activated more broadly during development (Kask *et al.*, 2000). Likewise, expression of the α-CaMKII promoter has been reported in a few studies to be restricted to area CA1 in 3-week old mice (Tsien *et al.*, 1996a), whereas expression is more widespread in adults (Nakazawa *et al.*, 2004). Thus, one should be careful to evaluate deletion patterns as a function of age, especially at the ages used for behavioral characterization.

As mentioned previously, fairly restricted expression of transgenes can be obtained using distinct promoters. For many brain regions however, no specific promoters are available or currently known. For example, no known promoters restrict gene expression to the amygdala. One way to spatially refine gene manipulation is by using two different promoters of which the expression pattern is different, but partly overlapping. As with many techniques that have been used in the study of mouse memory formation, inspiration can be found in work on *Drosophila* memory. For instance, Suster *et al.* (2004) used the combinatorial Gal4/Gal80 transgenic system to achieve ultra-precise transgene expression. The Gal4 transcriptional activator was expressed specifically in cholinergic neurons and an inhibitor of Gal4, Gal80, was expressed in a distinct but overlapping subset of neurons. This approach enabled them to reduce the expression of Gal4 from 200 to 20 neurons. Similar combinations could be made in mice by using one promoter to drive Cre expression and a second promoter to drive a construct that contains a floxed STOP codon prior to the transgene. Using this method, the STOP codon would be excised only in those regions that express both Cre and the floxed STOP-transgene constructs. This strategy will become even more enticing as a greater number of specific promoters and Cre-expressing lines become available (see for instance the GENSAT project).

The use of Cre/loxP recombination allows for more spatially restricted manipulation of genes bypassing lethal or severe developmental defects resulting from the loss of a gene in the entire organism. However, the use of Cre recombinase is not without any consequences. Several *in vitro* studies have indicated that high levels of Cre expression alone can cause reduced growth, cytopathic effects, and even damage DNA (Loonstra et al., 2001; Pfeifer et al., 2001). More recently, a study by Forni and colleagues (2006) indicated that high levels of Cre expression *in vivo* in neuronal progenitor cells hampers brain development leading to microencephaly (the abnormal smallness of the head) and hydrocephaly (the accumulation of cerebrospinal fluid in the cavities and ventricles of the brain). These are clearly confounding factors that can result in learning-unrelated side effects.

An alternative method to obtain high spatial resolution is by using a virus to drive transgene expression. Rumpel and colleagues (2005) used a viral approach to examine the role of AMPA receptor trafficking in relation to learning and memory. They injected a vector that encodes the GluR1 subunit fused with GFP into the mouse amygdala. Their study showed that learning an association between a tone and the delivery of a foot shock drives GluR1-containing AMPA receptors into the synapses of a large fraction of postsynaptic neurons in the lateral amygdala. Importantly, this study indicated the potency of using viral transgene delivery to study the role of specific genes.

The tTA and rtTA system allows researchers to manipulate gene function in a conditional and reversible manner, but it should be noted that dox itself can have several side effects on behavior. Riond and Riviere (1988) showed that long-term administration of dox creates an imbalance of the intestinal flora, resulting in diarrhea and some cases even cause colitis (the inflammation of the colon). So it is important to include the relevant control group fed dox, but lacking the tetO-regulated transgene.

Another issue that makes the interpretation of gene function difficult is the confounding effect of genetic background on behavioral phenotypes (Bucan and Abel, 2002; Crawley et al., 1997). The deletion of two isoforms of the transcription factor CREB may result in spatial learning and memory impairments in the Morris water maze on one strain of mice, but not another (Bourtchuladze et al., 1994; Gass et al., 1998; Graves et al., 2002; Kogan et al., 1997).

Also environmental influences can modify behavioral phenotypes. Crabbe et al. (1999) tested several frequently used mouse lines in three different laboratories at the exact same time of day. They subjected the different lines to several behavioral paradigms. Despite the exact same background and task parameters being used by access laboratories, behavioral findings varied significantly among the different laboratories involved.

One major challenge in the field of mammalian research is to determine which genes are really needed in which specific brain structure or system. Traditionally, a "necessity approach" is used: in which brain regions does the loss of a particular gene result in memory deficits? An alternative strategy that has frequently been used to determine the role of specific genes in memory formation is a so-called "sufficiency approach" examining in which brain region or regions rescue of the function of a specific gene is sufficient to rescue memory deficits. This approach has begun to be developed in *Drosophila*, but is largely absent from mouse experiments. By first knocking out *rutabaga* everywhere, and then selectively expressing the wild-type *rutabaga* adenylyl cyclase specifically in the kenyon cells of mushroom bodies of the *Drosophila* brain, Zars *et al.* (2000) showed that this restricted expression of the *rutabaga* adenylyl cyclase is sufficient for the formation of olfactory short-term memories. This is the sort of powerful demonstration of function and site of function that mouse behavioral genetics is currently lacking.

Despite the concerns and limitations, the temporal and spatial resolution achieved in genetic approaches to mouse memory over the past decade have been critical for refining our understanding of the molecular and cellular basis of memory. A major remaining challenge is to refine these temporal and spatial resolutions. Combinatorial approaches using distinct, refined promoters to drive multiple components, such as combining promoter1-floxed Stop-tTA transgenes with promoter2-cre, provide one potential avenue for spatial refinement, as does a similar strategy using viral injection of Cre-expressing constructs in mice in which a broad promoter drives a construct in which a transgene is preceded by a STOP codon flanked with loxP sites.

The wealth of Cre lines being developed as part of the GENSAT BAC transgenic project (http://www.gensat.org) should soon provide an expanded repertoire of such region-specific Cre lines and the use of tamoxifen-inducible Cre for many of these BAC lines will expand their usefulness. This will be especially true if the GENSAT lines unearth more interesting activity-regulated promoters with more selective induction patterns than, for example, the immediate early gene *c-fos*. The temporal resolution achieved with light-activated channels is ideal, but this strategy only addresses cellular activity not biochemistry. The development of biochemical manipulations that could parallel this temporal precision would be a dramatic advance for the study of memory processes guided by biochemical cascades that function on timescales of seconds to minutes.

Acknowledgments

We thank Christopher Vecsey and Joshua Hawk for comments on a previous version of this manuscript. This work was supported by NIMH Grants R01 MH60244, and P50MH6404501 (Project 4 to T.A.; R. Gur, Conte Center P.I.) and by a fellowship from the Netherlands Organization for Scientific Research (NWO, Rubicon Grant: 825.07.29 to R.H.).

References

Abel, T., and Lattal, K. M. (2001). Molecular mechanisms of memory acquisition, consolidation and retrieval. *Curr. Opin. Neurobiol.* **11,** 180–187.

Abel, T., Nguyen, P. V., Barad, M., Deuel, T. A., Kandel, E. R., and Bourtchouladze, R. (1997). Genetic demonstration of a role for PKA in the late phase of LTP and in hippocampus-based long-term memory. *Cell* **88,** 615–626.

Abraham, W. C., Mason, S. E., Demmer, J., Williams, J. M., Richardson, C. L., Tate, W. P., Lawlor, P. A., and Dragunow, M. (1993). Correlations between immediate early gene induction and the persistence of long-term potentiation. *Neuroscience* **56,** 717–727.

Adamantidis, A. R., Zhang, F., Aravanis, A. M., Deisseroth, K., and de Lecea, L. (2007). Neural substrates of awakening probed with optogenetic control of hypocretin neurons. *Nature* **450,** 420–424.

Avissar, S., Barki-Harrington, L., Nechamkin, Y., Roitman, G., and Schreiber, G. (2001). Elevated dopamine receptor-coupled G(s) protein measures in mononuclear leukocytes of patients with schizophrenia. *Schizophr. Res.* **47,** 37–47.

Bechara, A., Tranel, D., Damasio, H., Adolphs, R., Rockland, C., and Damasio, A. R. (1995). Double dissociation of conditioning and declarative knowledge relative to the amygdala and hippocampus in humans. *Science* **269,** 1115–1118.

Blendy, J. A., Kaestner, K. H., Schmid, W., Gass, P., and Schutz, G. (1996). Targeting of the CREB gene leads to up-regulation of a novel CREB mRNA isoform. *EMBO J.* **15,** 1098–1106.

Bliss, T. V., and Collingridge, G. L. (1993). A synaptic model of memory: Long-term potentiation in the hippocampus. *Nature* **361,** 31–39.

Bourtchuladze, R., Frenguelli, B., Blendy, J., Cioffi, D., Schutz, G., and Silva, A. J. (1994). Deficient long-term memory in mice with a targeted mutation of the cAMP-responsive element-binding protein. *Cell* **79,** 59–68.

Boyden, E. S., Zhang, F., Bamberg, E., Nagel, G., and Deisseroth, K. (2005). Millisecond-timescale, genetically targeted optical control of neural activity. *Nat. Neurosci.* **8,** 1263–1268.

Branda, C. S., and Dymecki, S. M. (2004). Talking about a revolution: The impact of site-specific recombinases on genetic analyses in mice. *Dev. Cell* **6,** 7–28.

Brandon, E. P., Zhuo, M., Huang, Y. Y., Qi, M., Gerhold, K. A., Burton, K. A., Kandel, E. R., McKnight, G. S., and Idzerda, R. L. (1995). Hippocampal long-term depression and depotentiation are defective in mice carrying a targeted disruption of the gene encoding the RI beta subunit of cAMP-dependent protein kinase. *Proc. Natl. Acad. Sci. USA* **92,** 8851–8855.

Brenner, M., Kisseberth, W. C., Su, Y., Besnard, F., and Messing, A. (1994). GFAP promoter directs astrocyte-specific expression in transgenic mice. *J. Neurosci.* **14,** 1030–1037.

Brocard, J., Warot, X., Wendling, O., Messaddeq, N., Vonesch, J. L., Chambon, P., and Metzger, D. (1997). Spatio-temporally controlled site-specific somatic mutagenesis in the mouse. *Proc. Natl. Acad. Sci. USA* **94,** 14559–14563.

Brockschnieder, D., Lappe-Siefke, C., Goebbels, S., Boesl, M. R., Nave, K. A., and Riethmacher, D. (2004). Cell depletion due to diphtheria toxin fragment A after Cre-mediated recombination. *Mol. Cell. Biol.* **24,** 7636–7642.

Bucan, M., and Abel, T. (2002). The mouse: Genetics meets behaviour. *Nat. Rev. Genet.* **3,** 114–123.

Cao, X., Wang, H., Mei, B., An, S., Yin, L., Wang, L. P., and Tsien, J. Z. (2008). Inducible and selective erasure of memories in the mouse brain via chemical-genetic manipulation. *Neuron* **60,** 353–366.

Caroni, P. (1997). Overexpression of growth-associated proteins in the neurons of adult transgenic mice. *J. Neurosci. Methods* **71,** 3–9.

Casper, K. B., Jones, K., and McCarthy, K. D. (2007). Characterization of astrocyte-specific conditional knockouts. *Genesis* **45**, 292–299.

Chalfie, M., Tu, Y., Euskirchen, G., Ward, W. W., and Prasher, D. C. (1994). Green fluorescent protein as a marker for gene expression. *Science* **263**, 802–805.

Chen, C., Rainnie, D. G., Greene, R. W., and Tonegawa, S. (1994). Abnormal fear response and aggressive behavior in mutant mice deficient for alpha-calcium-calmodulin kinase II. *Science* **266**, 291–294.

Cheng, L., Jin, Z., Liu, L., Yan, Y., Li, T., Zhu, X., and Jing, N. (2004). Characterization and promoter analysis of the mouse nestin gene. *FEBS Lett.* **565**, 195–202.

Clegg, C. H., Correll, L. A., Cadd, G. G., and McKnight, G. S. (1987). Inhibition of intracellular cAMP-dependent protein kinase using mutant genes of the regulatory type I subunit. *J. Biol. Chem.* **262**, 13111–13119.

Collingridge, G. L., Kehl, S. J., and McLennan, H. (1983). Excitatory amino acids in synaptic transmission in the Schaffer collateral-commissural pathway of the rat hippocampus. *J. Physiol.* **334**, 33–46.

Conklin, B. R., Hsiao, E. C., Claeysen, S., Dumuis, A., Srinivasan, S., Forsayeth, J. R., Guettier, J. M., Chang, W. C., Pei, Y., McCarthy, K. D., Nissenson, R. A., Wess, J., et al. (2008). Engineering GPCR signaling pathways with RASSLs. *Nat. Methods* **5**, 673–678.

Coons, A. H., Creech, H. J., and Jones, R. N. (1941). Immunological properties of an antibody containing a fluorescent group. *Proc. Soc. Exp. Biol. Med.* **47**, 200–202.

Crabbe, J. C., Wahlsten, D., and Dudek, B. C. (1999). Genetics of mouse behavior: Interactions with laboratory environment. *Science* **284**, 1670–1672.

Crawley, J. N., Belknap, J. K., Collins, A., Crabbe, J. C., Frankel, W., Henderson, N., Hitzemann, R. J., Maxson, S. C., Miner, L. L., Silva, A. J., Wehner, J. M., Wynshaw-Boris, A., et al. (1997). Behavioral phenotypes of inbred mouse strains: Implications and recommendations for molecular studies. *Psychopharmacology (Berl)* **132**, 107–124.

de Lecea, L., Kilduff, T. S., Peyron, C., Gao, X., Foye, P. E., Danielson, P. E., Fukuhara, C., Battenberg, E. L., Gautvik, V. T., Bartlett, F. S., II, Frankel, W. N., van den Pol, A. N., et al. (1998). The hypocretins: Hypothalamus-specific peptides with neuroexcitatory activity. *Proc. Natl. Acad. Sci. USA* **95**, 322–327.

Deutsch, J. A. (1993). Spatial learning in mutant mice. *Science* **262**, 760–763.

Duffy, S. N., and Nguyen, P. V. (2003). Postsynaptic application of a peptide inhibitor of cAMP-dependent protein kinase blocks expression of long-lasting synaptic potentiation in hippocampal neurons. *J. Neurosci.* **23**, 1142–1150.

Erdmann, G., Schutz, G., and Berger, S. (2007). Inducible gene inactivation in neurons of the adult mouse forebrain. *BMC Neurosci.* **8**, 63.

Feil, R., Brocard, J., Mascrez, B., LeMeur, M., Metzger, D., and Chambon, P. (1996). Ligand-activated site-specific recombination in mice. *Proc. Natl. Acad. Sci. USA* **93**, 10887–10890.

Feil, R., Wagner, J., Metzger, D., and Chambon, P. (1997). Regulation of Cre recombinase activity by mutated estrogen receptor ligand-binding domains. *Biochem. Biophys. Res. Commun.* **237**, 752–757.

Feng, G., Mellor, R. H., Bernstein, M., Keller-Peck, C., Nguyen, Q. T., Wallace, M., Nerbonne, J. M., Lichtman, J. W., and Sanes, J. R. (2000). Imaging neuronal subsets in transgenic mice expressing multiple spectral variants of GFP. *Neuron* **28**, 41–51.

Fischer, M., Rulicke, T., Raeber, A., Sailer, A., Moser, M., Oesch, B., Brandner, S., Aguzzi, A., and Weissmann, C. (1996). Prion protein (PrP) with amino-proximal deletions restoring susceptibility of PrP knockout mice to scrapie. *EMBO J.* **15**, 1255–1264.

Forni, P. E., Scuoppo, C., Imayoshi, I., Taulli, R., Dastru, W., Sala, V., Betz, U. A., Muzzi, P., Martinuzzi, D., Vercelli, A. E., Kageyama, R., and Ponzetto, C. (2006). High levels of Cre expression in neuronal progenitors cause defects in brain development leading to microencephaly and hydrocephaly. *J. Neurosci.* **26**, 9593–9602.

Forrest, D., Yuzaki, M., Soares, H. D., Ng, L., Luk, D. C., Sheng, M., Stewart, C. L., Morgan, J. I., Connor, J. A., and Curran, T. (1994). Targeted disruption of NMDA receptor 1 gene abolishes NMDA response and results in neonatal death. *Neuron* **13,** 325–338.

Forss-Petter, S., Danielson, P. E., Catsicas, S., Battenberg, E., Price, J., Nerenberg, M., and Sutcliffe, J. G. (1990). Transgenic mice expressing beta-galactosidase in mature neurons under neuron-specific enolase promoter control. *Neuron* **5,** 187–197.

Frey, U., Huang, Y. Y., and Kandel, E. R. (1993). Effects of cAMP simulate a late stage of LTP in hippocampal CA1 neurons. *Science* **260,** 1661–1664.

Frey, U., Frey, S., Schollmeier, F., and Krug, M. (1996). Influence of actinomycin D, a RNA synthesis inhibitor, on long-term potentiation in rat hippocampal neurons *in vivo* and *in vitro. J. Physiol.* **490** (Pt 3), 703–711.

Furth, P. A., St Onge, L., Boger, H., Gruss, P., Gossen, M., Kistner, A., Bujard, H., and Hennighausen, L. (1994). Temporal control of gene expression in transgenic mice by a tetracycline-responsive promoter. *Proc. Natl. Acad. Sci. USA* **91,** 9302–9306.

Fuss, B., Afshari, F. S., Colello, R. J., and Macklin, W. B. (2001). Normal CNS myelination in transgenic mice overexpressing MHC class I H-2L(d) in oligodendrocytes. *Mol. Cell. Neurosci.* **18,** 221–234.

Gass, P., Wolfer, D. P., Balschun, D., Rudolph, D., Frey, U., Lipp, H. P., and Schutz, G. (1998). Deficits in memory tasks of mice with CREB mutations depend on gene dosage. *Learn Mem.* **5,** 274–288.

Genoux, D., Haditsch, U., Knobloch, M., Michalon, A., Storm, D., and Mansuy, I. M. (2002). Protein phosphatase 1 is a molecular constraint on learning and memory. *Nature* **418,** 970–975.

Ginty, D. D., Kornhauser, J. M., Thompson, M. A., Bading, H., Mayo, K. E., Takahashi, J. S., and Greenberg, M. E. (1993). Regulation of CREB phosphorylation in the suprachiasmatic nucleus by light and a circadian clock. *Science* **260,** 238–241.

Gosgnach, S., Lanuza, G. M., Butt, S. J., Saueressig, H., Zhang, Y., Velasquez, T., Riethmacher, D., Callaway, E. M., Kiehn, O., and Goulding, M. (2006). V1 spinal neurons regulate the speed of vertebrate locomotor outputs. *Nature* **440,** 215–219.

Gossen, M., Bonin, A. L., Freundlieb, S., and Bujard, H. (1994). Inducible gene expression systems for higher eukaryotic cells. *Curr. Opin. Biotechnol.* **5,** 516–520.

Gossen, M., Freundlieb, S., Bender, G., Muller, G., Hillen, W., and Bujard, H. (1995). Transcriptional activation by tetracyclines in mammalian cells. *Science* **268,** 1766–1769.

Grant, S. G., O'Dell, T. J., Karl, K. A., Stein, P. L., Soriano, P., and Kandel, E. R. (1992). Impaired long-term potentiation, spatial learning, and hippocampal development in fyn mutant mice. *Science* **258,** 1903–1910.

Graves, L., Dalvi, A., Lucki, I., Blendy, J. A., and Abel, T. (2002). Behavioral analysis of CREB alphadelta mutation on a B6/129 F1 hybrid background. *Hippocampus* **12,** 18–26.

Gross, C., Zhuang, X., Stark, K., Ramboz, S., Oosting, R., Kirby, L., Santarelli, L., Beck, S., and Hen, R. (2002). Serotonin1A receptor acts during development to establish normal anxiety-like behaviour in the adult. *Nature* **416,** 396–400.

Gu, H., Marth, J. D., Orban, P. C., Mossmann, H., and Rajewsky, K. (1994). Deletion of a DNA polymerase beta gene segment in T cells using cell type-specific gene targeting. *Science* **265,** 103–106.

Halassa, M. M., Florian, C., Fellin, T., Munoz, J. R., Lee, S. Y., Abel, T., Haydon, P. G., and Frank, M. G. (2009). Astrocytic modulation of sleep homeostasis and cognitive consequences of sleep loss. *Neuron* **61,** 213–219.

Harvey, C. D., and Svoboda, K. (2007). Locally dynamic synaptic learning rules in pyramidal neuron dendrites. *Nature* **450,** 1195–1200.

Havekes, R., Nijholt, I. M., Visser, A. K., Eisel, U. L., and Van der Zee, E. A. (2008). Transgenic inhibition of neuronal calcineurin activity in the forebrain facilitates fear conditioning, but inhibits the extinction of contextual fear memories. *Neurobiol. Learn Mem.* **89,** 595–598.

Hebb, D. O. (1949). The Organization of Behavior Wiley, New York.

Heim, R., and Tsien, R. Y. (1996). Engineering green fluorescent protein for improved brightness, longer wavelengths and fluorescence resonance energy transfer. *Curr. Biol.* **6,** 178–182.

Heim, R., Prasher, D. C., and Tsien, R. Y. (1994). Wavelength mutations and posttranslational autoxidation of green fluorescent protein. *Proc. Natl. Acad. Sci. USA* **91,** 12501–12504.

Heim, R., Cubitt, A. B., and Tsien, R. Y. (1995). Improved green fluorescence. *Nature* **373,** 663–664.

Huang, Y. Y., and Kandel, E. R. (1994). Recruitment of long-lasting and protein kinase A-dependent long-term potentiation in the CA1 region of hippocampus requires repeated tetanization. *Learn Mem.* **1,** 74–82.

Imayoshi, I., Ohtsuka, T., Metzger, D., Chambon, P., and Kageyama, R. (2006). Temporal regulation of Cre recombinase activity in neural stem cells. *Genesis* **44,** 233–238.

Imayoshi, I., Sakamoto, M., Ohtsuka, T., Takao, K., Miyakawa, T., Yamaguchi, M., Mori, K., Ikeda, T., Itohara, S., and Kageyama, R. (2008). Roles of continuous neurogenesis in the structural and functional integrity of the adult forebrain. *Nat. Neurosci.* **11,** 1153–1161.

Isiegas, C., Park, A., Kandel, E. R., Abel, T., and Lattal, K. M. (2006). Transgenic inhibition of neuronal protein kinase A activity facilitates fear extinction. *J. Neurosci.* **26,** 12700–12707.

Isiegas, C., McDonough, C., Huang, T., Havekes, R., Fabian, S., Wu, L. J., Xu, H., Zhao, M. G., Kim, J. I., Lee, Y. S., Lee, H. R., Ko, H. G., *et al.* (2008). A novel conditional genetic system reveals that increasing neuronal cAMP enhances memory and retrieval. *J. Neurosci.* **28,** 6220–6230.

Karpova, A. Y., Tervo, D. G., Gray, N. W., and Svoboda, K. (2005). Rapid and reversible chemical inactivation of synaptic transmission in genetically targeted neurons. *Neuron* **48,** 727–735.

Kask, K., Jerecic, J., Zamanillo, D., Wilbertz, J., Sprengel, R., and Seeburg, P. H. (2000). Developmental profile of kainate receptor subunit KA1 revealed by Cre expression in YAC transgenic mice. *Brain Res.* **876,** 55–61.

Kelly, M. P., Cheung, Y. F., Favilla, C., Siegel, S. J., Kanes, S. J., Houslay, M. D., and Abel, T. (2008). Constitutive activation of the G-protein subunit Galphas within forebrain neurons causes PKA-dependent alterations in fear conditioning and cortical Arc mRNA expression. *Learn Mem.* **15,** 75–83.

Kida, S., Josselyn, S. A., de Ortiz, S. P., Kogan, J. H., Chevere, I., Masushige, S., and Silva, A. J. (2002). CREB required for the stability of new and reactivated fear memories. *Nat. Neurosci.* **5,** 348–355.

Kogan, J. H., Frankland, P. W., Blendy, J. A., Coblentz, J., Marowitz, Z., Schutz, G., and Silva, A. J. (1997). Spaced training induces normal long-term memory in CREB mutant mice. *Curr. Biol.* **7,** 1–11.

LeDoux, J. E. (2000). Emotion circuits in the brain. *Annu. Rev. Neurosci.* **23,** 155–184.

Lerchner, W., Xiao, C., Nashmi, R., Slimko, E. M., van Trigt, L., Lester, H. A., and Anderson, D. J. (2007). Reversible silencing of neuronal excitability in behaving mice by a genetically targeted, ivermectin-gated Cl- channel. *Neuron* **54,** 35–49.

Li, W., Zhou, Y., Jentsch, J. D., Brown, R. A., Tian, X., Ehninger, D., Hennah, W., Peltonen, L., Lonnqvist, J., Huttunen, M. O., Kaprio, J., Trachtenberg, J. T., *et al.* (2007). Specific developmental disruption of disrupted-in-schizophrenia-1 function results in schizophrenia-related phenotypes in mice. *Proc. Natl. Acad. Sci. USA* **104,** 18280–18285.

Livet, J., Weissman, T. A., Kang, H., Draft, R. W., Lu, J., Bennis, R. A., Sanes, J. R., and Lichtman, J. W. (2007). Transgenic strategies for combinatorial expression of fluorescent proteins in the nervous system. *Nature* **450,** 56–62.

Loonstra, A., Vooijs, M., Beverloo, H. B., Allak, B. A., van Drunen, E., Kanaar, R., Berns, A., and Jonkers, J. (2001). Growth inhibition and DNA damage induced by Cre recombinase in mammalian cells. *Proc. Natl. Acad. Sci. USA* **98,** 9209–9214.

Malleret, G., Haditsch, U., Genoux, D., Jones, M. W., Bliss, T. V., Vanhoose, A. M., Weitlauf, C., Kandel, E. R., Winder, D. G., and Mansuy, I. M. (2001). Inducible and reversible enhancement of learning, memory, and long-term potentiation by genetic inhibition of calcineurin. *Cell* **104**, 675–686.

Mansuy, I. M., Winder, D. G., Moallem, T. M., Osman, M., Mayford, M., Hawkins, R. D., and Kandel, E. R. (1998). Inducible and reversible gene expression with the rtTA system for the study of memory. *Neuron* **21**, 257–265.

Martin, S. J., Grimwood, P. D., and Morris, R. G. (2000). Synaptic plasticity and memory: An evaluation of the hypothesis. *Annu. Rev. Neurosci.* **23**, 649–711.

Matsuo, N., Reijmers, L., and Mayford, M. (2008). Spine-type-specific recruitment of newly synthesized AMPA receptors with learning. *Science* **319**, 1104–1147.

Mayford, M., Bach, M. E., Huang, Y. Y., Wang, L., Hawkins, R. D., and Kandel, E. R. (1996). Control of memory formation through regulated expression of a CaMKII transgene. *Science* **274**, 1678–1683.

Mayr, B., and Montminy, M. (2001). Transcriptional regulation by the phosphorylation-dependent factor CREB. *Nat. Rev. Mol. Cell Biol.* **2**, 599–609.

McHugh, T. J., Blum, K. I., Tsien, J. Z., Tonegawa, S., and Wilson, M. A. (1996). Impaired hippocampal representation of space in CA1-specific NMDAR1 knockout mice. *Cell* **87**, 1339–1349.

McHugh, T. J., Jones, M. W., Quinn, J. J., Balthasar, N., Coppari, R., Elmquist, J. K., Lowell, B. B., Fanselow, M. S., Wilson, M. A., and Tonegawa, S. (2007). Dentate gyrus NMDA receptors mediate rapid pattern separation in the hippocampal network. *Science* **317**, 94–99.

Memo, M., Kleinman, J. E., and Hanbauer, I. (1983). Coupling of dopamine D1 recognition sites with adenylate cyclase in nuclei accumbens and caudatus of schizophrenics. *Science* **221**, 1304–1307.

Metzger, D., Ali, S., Bornert, J. M., and Chambon, P. (1995a). Characterization of the amino-terminal transcriptional activation function of the human estrogen receptor in animal and yeast cells. *J. Biol. Chem.* **270**, 9535–9542.

Metzger, D., Clifford, J., Chiba, H., and Chambon, P. (1995b). Conditional site-specific recombination in mammalian cells using a ligand-dependent chimeric Cre recombinase. *Proc. Natl. Acad. Sci. USA* **92**, 6991–6995.

Minoretti, P., Politi, P., Coen, E., Di Vito, C., Bertona, M., Bianchi, M., and Emanuele, E. (2006). The T393C polymorphism of the GNAS1 gene is associated with deficit schizophrenia in an Italian population sample. *Neurosci. Lett.* **397**, 159–163.

Moriyoshi, K., Masu, M., Ishii, T., Shigemoto, R., Mizuno, N., and Nakanishi, S. (1991). Molecular cloning and characterization of the rat NMDA receptor. *Nature* **354**, 31–37.

Morris, R. G., Anderson, E., Lynch, G. S., and Baudry, M. (1986). Selective impairment of learning and blockade of long-term potentiation by an N-methyl-D-aspartate receptor antagonist, AP5. *Nature* **319**, 774–776.

Nagel, G., Ollig, D., Fuhrmann, M., Kateriya, S., Musti, A. M., Bamberg, E., and Hegemann, P. (2002). Channelrhodopsin-1: A light-gated proton channel in green algae. *Science* **296**, 2395–2398.

Nagel, G., Szellas, T., Huhn, W., Kateriya, S., Adeishvili, N., Berthold, P., Ollig, D., Hegemann, P., and Bamberg, E. (2003). Channelrhodopsin-2, a directly light-gated cation-selective membrane channel. *Proc. Natl. Acad. Sci. USA* **100**, 13940–13945.

Nakanishi, S. (1992). Molecular diversity of glutamate receptors and implications for brain function. *Science* **258**, 597–603.

Nakashiba, T., Young, J. Z., McHugh, T. J., Buhl, D. L., and Tonegawa, S. (2008). Transgenic inhibition of synaptic transmission reveals role of CA3 output in hippocampal learning. *Science* **319**, 1260–1264.

Nakazawa, K., Quirk, M. C., Chitwood, R. A., Watanabe, M., Yeckel, M. F., Sun, L. D., Kato, A., Carr, C. A., Johnston, D., Wilson, M. A., and Tonegawa, S. (2002). Requirement for hippocampal CA3 NMDA receptors in associative memory recall. *Science* 297, 211–218.

Nakazawa, K., Sun, L. D., Quirk, M. C., Rondi-Reig, L., Wilson, M. A., and Tonegawa, S. (2003). Hippocampal CA3 NMDA receptors are crucial for memory acquisition of one-time experience. *Neuron* 38, 305–315.

Nakazawa, K., McHugh, T. J., Wilson, M. A., and Tonegawa, S. (2004). NMDA receptors, place cells and hippocampal spatial memory. *Nat. Rev. Neurosci.* 5, 361–372.

Nguyen, P. V., and Woo, N. H. (2003). Regulation of hippocampal synaptic plasticity by cyclic AMP-dependent protein kinases. *Prog. Neurobiol.* 71, 401–437.

Oberdick, J., Smeyne, R. J., Mann, J. R., Zackson, S., and Morgan, J. I. (1990). A promoter that drives transgene expression in cerebellar Purkinje and retinal bipolar neurons. *Science* 248, 223–226.

O'Keefe, J., and Dostrovsky, J. (1971). The hippocampus as a spatial map. Preliminary evidence from unit activity in the freely-moving rat. *Brain Res.* 34, 171–175.

Palmiter, R. D., and Brinster, R. L. (1986). Germ-line transformation of mice. *Annu. Rev. Genet.* 20, 465–499.

Palmiter, R. D., Brinster, R. L., Hammer, R. E., Trumbauer, M. E., Rosenfeld, M. G., Birnberg, N. C., and Evans, R. M. (1982). Dramatic growth of mice that develop from eggs microinjected with metallothionein-growth hormone fusion genes. *Nature* 300, 611–615.

Pascual, O., Casper, K. B., Kubera, C., Zhang, J., Revilla-Sanchez, R., Sul, J. Y., Takano, H., Moss, S. J., McCarthy, K., and Haydon, P. G. (2005). Astrocytic purinergic signaling coordinates synaptic networks. *Science* 310, 113–116.

Peyron, C., Tighe, D. K., van den Pol, A. N., de Lecea, L., Heller, H. C., Sutcliffe, J. G., and Kilduff, T. S. (1998). Neurons containing hypocretin (orexin) project to multiple neuronal systems. *J. Neurosci.* 18, 9996–10015.

Pfeifer, A., Brandon, E. P., Kootstra, N., Gage, F. H., and Verma, I. M. (2001). Delivery of the Cre recombinase by a self-deleting lentiviral vector: Efficient gene targeting *in vivo*. *Proc. Natl. Acad. Sci. USA* 98, 11450–11455.

Prasher, D. C., Eckenrode, V. K., Ward, W. W., Prendergast, F. G., and Cormier, M. J. (1992). Primary structure of the *Aequorea victoria* green-fluorescent protein. *Gene* 111, 229–233.

Ramos-Vara, J. A. (2005). Technical aspects of immunohistochemistry. *Vet. Pathol.* 42, 405–426.

Reijmers, L. G., Coats, J. K., Pletcher, M. T., Wiltshire, T., Tarantino, L. M., and Mayford, M. (2006). A mutant mouse with a highly specific contextual fear-conditioning deficit found in an N-ethyl-N-nitrosourea (ENU) mutagenesis screen. *Learn Mem.* 13, 143–149.

Reijmers, L. G., Perkins, B. L., Matsuo, N., and Mayford, M. (2007). Localization of a stable neural correlate of associative memory. *Science* 317, 1230–1233.

Riond, J. L., and Riviere, J. E. (1988). Pharmacology and toxicology of doxycycline. *Vet. Hum. Toxicol.* 30, 431–443.

Rotenberg, A., Abel, T., Hawkins, R. D., Kandel, E. R., and Muller, R. U. (2000). Parallel instabilities of long-term potentiation, place cells, and learning caused by decreased protein kinase A activity. *J. Neurosci.* 20, 8096–8102.

Rumpel, S., LeDoux, J., Zador, A., and Malinow, R. (2005). Postsynaptic receptor trafficking underlying a form of associative learning. *Science* 308, 83–88.

Sauer, B., and Henderson, N. (1988). Site-specific DNA recombination in mammalian cells by the Cre recombinase of bacteriophage P1. *Proc. Natl. Acad. Sci. USA* 85, 5166–5170.

Scharf, M. T., Woo, N. H., Lattal, K. M., Young, J. Z., Nguyen, P. V., and Abel, T. (2002). Protein synthesis is required for the enhancement of long-term potentiation and long-term memory by spaced training. *J. Neurophysiol.* 87, 2770–2777.

Shaner, N. C., Campbell, R. E., Steinbach, P. A., Giepmans, B. N., Palmer, A. E., and Tsien, R. Y. (2004). Improved monomeric red, orange and yellow fluorescent proteins derived from *Discosoma* sp. red fluorescent protein. *Nat. Biotechnol.* 22, 1567–1572.

Silva, A. J., Paylor, R., Wehner, J. M., and Tonegawa, S. (1992a). Impaired spatial learning in alpha-calcium-calmodulin kinase II mutant mice. *Science* **257**, 206–211.

Silva, A. J., Stevens, C. F., Tonegawa, S., and Wang, Y. (1992b). Deficient hippocampal long-term potentiation in alpha-calcium-calmodulin kinase II mutant mice. *Science* **257**, 201–216.

Sineshchekov, O. A., Jung, K. H., and Spudich, J. L. (2002). Two rhodopsins mediate phototaxis to low- and high-intensity light in *Chlamydomonas reinhardtii*. *Proc. Natl. Acad. Sci. USA* **99**, 8689–8694.

Struthers, R. S., Vale, W. W., Arias, C., Sawchenko, P. E., and Montminy, M. R. (1991). Somatotroph hypoplasia and dwarfism in transgenic mice expressing a non-phosphorylatable CREB mutant. *Nature* **350**, 622–624.

Suster, M. L., Seugnet, L., Bate, M., and Sokolowski, M. B. (2004). Refining GAL4-driven transgene expression in Drosophila with a GAL80 enhancer-trap. *Genesis* **39**, 240–245.

Sweger, E. J., Casper, K. B., Scearce-Levie, K., Conklin, B. R., and McCarthy, K. D. (2007). Development of hydrocephalus in mice expressing the G(i)-coupled GPCR Ro1 RASSL receptor in astrocytes. *J. Neurosci.* **27**, 2309–2317.

Takahashi, J. S., Shimomura, K., and Kumar, V. (2008). Searching for genes underlying behavior: Lessons from circadian rhythms. *Science* **322**, 909–912.

Tan, E. M., Yamaguchi, Y., Horwitz, G. D., Gosgnach, S., Lein, E. S., Goulding, M., Albright, T. D., and Callaway, E. M. (2006). Selective and quickly reversible inactivation of mammalian neurons *in vivo* using the Drosophila allatostatin receptor. *Neuron* **51**, 157–170.

Tanaka, J., Horiike, Y., Matsuzaki, M., Miyazaki, T., Ellis-Davies, G. C., and Kasai, H. (2008). Protein synthesis and neurotrophin-dependent structural plasticity of single dendritic spines. *Science* **319**, 1683–1687.

Tsien, J. Z., Chen, D. F., Gerber, D., Tom, C., Mercer, E. H., Anderson, D. J., Mayford, M., Kandel, E. R., and Tonegawa, S. (1996a). Subregion- and cell type-restricted gene knockout in mouse brain. *Cell* **87**, 1317–1326.

Tsien, J. Z., Huerta, P. T., and Tonegawa, S. (1996b). The essential role of hippocampal CA1 NMDA receptor-dependent synaptic plasticity in spatial memory. *Cell* **87**, 1327–1338.

Vo, N., and Goodman, R. H. (2001). CREB-binding protein and p300 in transcriptional regulation. *J. Biol. Chem.* **276**, 13505–13508.

Wang, H., Feng, R., Phillip Wang, L., Li, F., Cao, X., and Tsien, J. Z. (2008). CaMKII activation state underlies synaptic labile phase of LTP and short-term memory formation. *Curr. Biol.* **18**, 1546–1554.

Woo, N. H., Duffy, S. N., Abel, T., and Nguyen, P. V. (2000). Genetic and pharmacological demonstration of differential recruitment of cAMP-dependent protein kinases by synaptic activity. *J. Neurophysiol.* **84**, 2739–2745.

Wood, M. A., Kaplan, M. P., Park, A., Blanchard, E. J., Oliveira, A. M., Lombardi, T. L., and Abel, T. (2005). Transgenic mice expressing a truncated form of CREB-binding protein (CBP) exhibit deficits in hippocampal synaptic plasticity and memory storage. *Learn Mem.* **12**, 111–119.

Wood, M. A., Attner, M. A., Oliveira, A. M., Brindle, P. K., and Abel, T. (2006). A transcription factor-binding domain of the coactivator CBP is essential for long-term memory and the expression of specific target genes. *Learn. Mem.* **13**, 609–617.

Wulff, P., Goetz, T., Leppa, E., Linden, A. M., Renzi, M., Swinny, J. D., Vekovischeva, O. Y., Sieghart, W., Somogyi, P., Korpi, E. R., Farrant, M., and Wisden, W. (2007). From synapse to behavior: Rapid modulation of defined neuronal types with engineered GABAA receptors. *Nat. Neurosci.* **10**, 923–929.

Zacharias, D. A., and Tsien, R. Y. (2006). Molecular biology and mutation of green fluorescent protein. *Methods Biochem. Anal.* **47**, 83–120.

Zars, T., Fischer, M., Schulz, R., and Heisenberg, M. (2000). Localization of a short-term memory in Drosophila. *Science* **288**, 672–675.

Zhang, Q., Pangrsic, T., Kreft, M., Krzan, M., Li, N., Sul, J. Y., Halassa, M., Van Bockstaele, E., Zorec, R., and Haydon, P. G. (2004). Fusion-related release of glutamate from astrocytes. *J. Biol. Chem.* **279**, 12724–12733.

2 Worm Watching: Imaging Nervous System Structure and Function in *Caenorhabditis elegans*

Jeremy Dittman

Department of Biochemistry, Weill Cornell Medical College, New York, NY, USA

I. Introduction
II. Real-Time Indicators of Neuronal Activity
  A. Properties of worm neurons
  B. Speed of electrical signaling
III. Monitoring Intracellular Calcium
  A. GECIs: Cameleon and G-CaMP
  B. Examples of GECIs in worm neurobiology
  C. Limitations of GECIs
IV. Imaging Presynaptic Terminals
  A. Quantifying presynaptic proteins
  B. pH-Sensitive GFP at the synapse
  C. Spatial localization as a signal
  D. Systematic profiling of presynaptic proteins
V. Imaging Postsynaptic Compartments
  A. Postsynaptic glutamate receptors
  B. Postsynaptic nicotinic acetylcholine receptors
  C. Other receptors
VI. Imaging Proxies for Neuropeptide Secretion
VII. Bimolecular Fluorescence Complementation in the Worm
  A. Combinatorial promoters
  B. GRASP
VIII. Methods for Immobilizing the Worm
  A. Pharmacological paralysis
  B. Glue

Advances in Genetics, Vol. 65
Copyright 2009, Elsevier Inc. All rights reserved.

0065-2660/09 $35.00
DOI: 10.1016/S0065-2660(09)65002-1

ABSTRACT

Caenorhabditis elegans has become a model system of choice for optical approaches to cellular biology largely due to its extraordinary combination of transparency, well-defined anatomy, rapid generation time, and simple genetics. In particular, studies in nervous system development and function have benefited tremendously since C. *elegans* was first examined under the microscope. After the introduction of green fluorescent protein as a means of following gene expression and protein localization in living animals, a variety of optical approaches have been developed for probing and perturbing neuronal activity. Microfluidic technologies have opened new possibilities for high-resolution imaging during behavior. Femtosecond pulsed lasers allow for precise severing of individual processes in the living animal. This chapter will cover some recent methodological advances in imaging worm neurons as well as some of the many biological details of the worm nervous system revealed by these new optical approaches. Advantages and limitations of these methods will be discussed in this chapter. © 2009, Elsevier Inc.

I. INTRODUCTION

One major goal of neuroscience is to connect the cellular activity of the nervous system with the resulting behavior of the animal. Numerous model systems, both vertebrate and invertebrate, have contributed greatly to our current understanding of nervous system function from the molecular to the systems level over the past century. One creature in particular has served as a model for neuronal function at the cross roads of genetics, circuits, and cell biology over the past 40 years since its introduction by Sydney Brenner in 1965 (Brenner, 1974). *Caenorhabditis elegans* is a small free-living soil nematode with a compact nervous system composed of 302 neurons making around 7000 chemical and electrical synapses. Despite the diminutive nature of its nervous system, C. *elegans* exhibits a host of distinct behaviors including various modes of locomotion, egg laying, mating behaviors, chemo- and thermotaxis, and rudimentary forms of learning (Bargmann and Kaplan, 1998; Chi et al., 2007; Giles et al., 2006; Kano et al., 2008; Mori et al., 2007; Schafer, 2005; Whittaker and Sternberg, 2004).

Forward genetic screens based on these behaviors have yielded thousands of genes that are crucial for the proper function of the worm nervous system and a large proportion of these genes have orthologs in the human genome, highlighting the great degree of molecular conservation within the bilaterian nervous system (Bargmann, 1998). In addition to revealing the set of genes that encode the nervous system, C. *elegans* was the first animal (and the only animal to date) to have its entire nervous system delineated at the electron micrographic level. John White and colleagues reconstructed the positions and identities of every neuron together with most of their chemical and electrical synapses, thereby providing a wiring diagram to accompany the neuronal gene set (White et al., 1986). Most neurons and their progenitor cells can be visualized in living animals using conventional light microscopy with Nomarski optics, and some of the first genes in nervous system development were found in the worm by observing alterations in the stereotypical progression of cell division, migration, and apoptosis that occurred in particular mutants. Over the past 30 years, lineage studies have revealed many of the developmental details behind the assembly of the worm nervous system (White et al., 1982). Electron and light microscopic studies in C. *elegans* exemplify a strategy of visualizing structures and processes in wild-type and mutant animals as a powerful means of learning about nervous system function. Another addition to these optical approaches was laser ablation, a technique that took advantage of the transparent cuticle and reproducible anatomy of the worm. A particular cell or its precursor was irradiated and killed with ultraviolet light thereby allowing the functions of a particular neuron to be assayed based on the consequences of its selective removal (Bargmann and Horvitz, 1991; Bargmann et al., 1993). The combination of genetics with both the detailed knowledge of neuronal wiring, and the impact of neuronal ablation has made possible detailed models of the neural circuits underlying behavior (Fig. 2.1).

Given the rich history of optical approaches in the neurobiology of C. *elegans* together with the relative ease by which transgenes can be expressed, it is no surprise that the worm was one of the first animals in which the jellyfish *Aequorea victoria* green fluorescent protein (GFP) was used as a fluorescent marker for *in vivo* visualization (Chalfie et al., 1994; Nonet, 1999). GFP rapidly became the reagent of choice for determining promoter expression patterns and protein localization (Boulin et al., 2006; Hobert and Loria, 2006). Furthermore, the diversity of genetically encoded fluorophores has grown exponentially over the past decade, making it possible to image in a variety of spectral channels, and

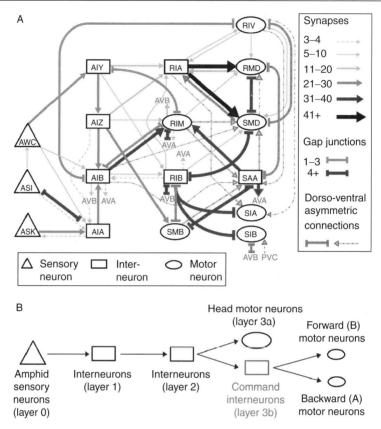

Figure 2.1. A predicted circuit for navigation. (A) Data from serial section reconstructions of electron micrographs were used to assemble a circuit. Each of the following neurons represents a bilaterally symmetric left–right pair: AWC, ASI, ASK, AIY, AIZ, AIB, AIA, RIA, RIM, RIB, and RIV. The head and neck motor neurons, SMD, SIA, SMB, and SIB, each have four members that innervate muscle quadrants. The interneuron SAA also has four members, a ventral and dorsal member on each side. RMD is a class of six radially arrayed neurons. Red dotted lines indicate connections that were asymmetric in the dorsoventral direction (e.g., seven of eight synapses from AIB to SMD are to the dorsal SMDs). The command interneurons are indicated in green. (B) A schematic showing information flow from sensory neurons to motor neurons. Reproduced from Gray et al. (2005). (See Page 2 in Color Plate Section at the end of the book.)

to use GFP derivatives as biosensors for ions such as calcium and protons, cAMP, inositol phospholipids, and a host of proteins (Miyawaki, 2003, 2005; Pologruto et al., 2004; Stauffer et al., 1998; Tsien, 1998; Zaccolo et al., 2000).

II. REAL-TIME INDICATORS OF NEURONAL ACTIVITY

When considering how to monitor neuronal activity with an optical probe, there are many definitions of activity on a wide variety of time scales. Historically, electrophysiologists recorded the voltage difference across the plasma membrane as the canonical measure of activity in excitable tissues such as neuron and muscle on a time scale of milliseconds to seconds. For neurons that support action potentials, the frequency of spikes has served as a useful definition of activity since spikes represent the output of a given neuron. However, subthreshold voltage changes driven by synaptic inputs and intrinsic conductances can also be of great interest in defining the activity of a neuron. In addition to voltage, other cellular parameters such as intracellular calcium have been very useful proxies for activity (Helmchen and Waters, 2002; Higley and Sabatini, 2008; Miyawaki, 2005). Many signal transduction pathways are coupled to membrane voltage via calcium influx so the accumulation of intracellular calcium can be an important parameter in defining the state of a neuron on a time scale of seconds to minutes (Hille, 2001). Longer term changes in the state of a neuron (minutes to hours) may be followed by monitoring the phosphorylation of particular proteins or by changes in gene expression (Greer and Greenberg, 2008). In addition to the range of time scales on which activity can be defined, spatially localized signaling can be a crucial aspect of neuronal activity. Neurons are morphologically complex cells and certain aspects of activity may be compartmentalized within the cell body, dendritic branch, or axon.

A. Properties of worm neurons

Given the variety of definitions for neuronal activity, it is important to consider which aspects of the nervous system are critical for a particular function or behavior. In the case of C. elegans, little is known about the detailed electrophysiological properties of many of its neurons due to their inaccessibility to conventional patch clamp and sharp electrode recording techniques. However, in the cases where successful neuronal recordings have been reported and in body wall muscle recordings, several differences from vertebrate neurons and muscle were apparent (Francis et al., 2003; Goodman et al., 1998; Jospin et al., 2002; Liu et al., 2005; Madison et al., 2005; Richmond and Jorgensen, 1999). The most striking difference was a general lack of fast sodium-dependent action potentials. Together with the conspicuous absence of a voltage-gated sodium channel ortholog in the C. elegans genome, it appears that the nervous system of this nematode does not operate in the currency of classical fast action potentials.

This does not exclude the role for fast calcium-dependent spiking or other forms of regenerative changes in membrane potential. Indeed, some neurons have been shown to display bistable potentials that depend on calcium

(Mellem *et al.*, 2008). Small deflections in membrane voltage can be amplified and stabilized at either a depolarized state (-20 mV) or a hyperpolarized state (-70 mV). Other neurons appear to have resting potentials only at the depolarized state, reminiscent of findings in the parasitic nematode *Ascaris*. In electrophysiological studies on *Ascaris lumbricoides* and *Ascaris suum*, where the large size of neurons allows for sharp electrode recordings, neurons and muscle appeared to sit at a relatively depolarized potential. The resting potential was maintained by basal sodium and chloride permeability rather than potassium permeability (Brading and Caldwell, 1971; Stretton *et al.*, 1992). *C. elegans* neurons are tiny with sparse branching (3 μm somas and 200–300 nm diameter neurites usually a few hundred microns in length). The small electrotonic structure coupled with low basal potassium conductance places a great importance on potassium channels since small changes in potassium channel conductance will rapidly impact the voltage across the entire neuron. Interestingly, *C. elegans* possesses approximately 70 genes encoding potassium channels; in particular, the class of 4TM-type potassium channels is greatly expanded in the worm genome relative to vertebrates (Salkoff *et al.*, 2005). Worm neurons appear to differentially express small subsets of these channel types so it is likely that the worm nervous system employs a rich repertoire of potassium channel-dependent voltage regulation (Salkoff *et al.*, 2005).

B. Speed of electrical signaling

Lack of sodium spikes does not equate with lack of speed in signaling. For a passive cable approximation of a neurite, the propagation speed (θ) of an electrical disturbance depends on the membrane and cytoplasmic resistivities (R_m and R_i) as well as membrane capacitance (C_m) and fiber diameter (d) according to the simple relation (Johnston and Wu, 1995):

$$\theta = \left(\frac{2d}{R_m R_i C_m^2} \right)^{1/2}$$

Given a typical *C. elegans* axon of 0.2 μm diameter with representative values for cytoplasmic resistivity and lipid bilayer capacitance, the propagation speed will be between 6 and 60 μm/ms depending on the membrane resistivity (here ranging between 10 and 1000 kΩ cm^2). These values correspond to space constants of between 300 and 3000 μm meaning that there is minimal attenuation of the voltage change along the length of a neuronal process. Even in the absence of precise information on the membrane resistivity, it is clear that in a matter of 10s of milliseconds, a signal could propagate across substantial fractions of the worm's length without requiring a regenerative process such as an action potential (Goodman *et al.*, 1998).

Despite the electrophysiological differences between vertebrate and nematode neurons, the fundamental process of chemical neurotransmission appears strikingly similar. Spontaneous acetylcholine (ACh) release recorded at the worm neuromuscular junction (NMJ) is highly reminiscent of its vertebrate counterparts in many respects. Miniature excitatory postsynaptic currents (mEPSCs) arrive stochastically with similar kinetics (Madison *et al.*, 2005; Richmond and Jorgensen, 1999; Richmond *et al.*, 1999). Evoked responses are highly calcium sensitive and a host of presynaptic proteins identified in worm have mammalian orthologs with similar function in synaptic transmission (Gracheva *et al.*, 2006; Koushika *et al.*, 2001; Madison *et al.*, 2005; McEwen *et al.*, 2006; Richmond and Jorgensen, 1999; Richmond *et al.*, 1999; Weimer *et al.*, 2003). Nevertheless, it is not known whether these synapses are utilized in a graded manner where analog voltage fluctuations modulate the rate of synaptic vesicle (SV) fusion, or if large calcium-dependent excursions in membrane voltage synchronize secretion in an all-or-none fashion.

III. MONITORING INTRACELLULAR CALCIUM

By far the most widely used optical indicators of neuronal activity are calcium dyes (Helmchen and Waters, 2002; Pologruto *et al.*, 2004). For most cell culture and brain slice imaging, fluorescent calcium indicators are either microinjected or loaded as cell-permeable acetoxymethyl esters into the target tissue and intracellular calcium concentration is inferred from the resulting fluorescence (Regehr and Tank, 1991, 1994). The ability to monitor this aspect of neuronal activity has greatly expanded our knowledge of processes underlying excitability, plasticity, and intracellular signaling (Greer and Greenberg, 2008; Helmchen and Waters, 2002). In some cases, conventional calcium dyes have been used successfully in *C. elegans* via direct injection (Dal Santo *et al.*, 1999). However, a major limitation of this approach is the lack of access and specificity: if the indicator cannot be directly and exclusively loaded into the tissue of interest, then the fluorescence is a complex mixture of calcium signals from various sources. Genetically encoded calcium indicators (GECIs) have become widely used indicators of neuronal activity in rodents, fish, flies, and worms because they bypass this limitation (Chalasani *et al.*, 2007; Higashijima *et al.*, 2003; Kerr and Schafer, 2006; Marella *et al.*, 2006; Miyawaki, 2005; Pologruto *et al.*, 2004).

A. GECIs: Cameleon and G-CaMP

A variety of GECIs have been designed for coupling calcium binding to alterations in fluorescence but the two major GECIs to be widely used in worms are cameleon and G-CaMP. Cameleon is a FRET-based ratiometric calcium sensor with the calcium binding domain of calmodulin and M13 (a calmodulin binding domain) flanked by CFP and YFP (Miyawaki, 2003, 2005). Upon calcium binding, changes in the distance and relative orientation of CFP and YFP increase the efficiency of resonance energy transfer such that YFP fluorescence increases while CFP fluorescence decreases while illuminating the sample in the CFP excitation spectrum ($\sim$440 nm). The fluorescence ratio is therefore a measure of the ambient calcium concentration, and the process of calculating the ratio has the added benefit of canceling out some of the artifacts associated with movement and light source fluctuations. G-CaMP is a single fluorophore generated from circularly permuted GFP flanked by calmodulin and the M13 peptide (Miyawaki, 2005; Nakai et al., 2001). Calcium binding directly influences the fluorophore so fluorescence can be related to calcium concentration in a straightforward manner.

B. Examples of GECIs in worm neurobiology

Most applications of calcium imaging in C. elegans thus far have been focused on its sensory physiology. Thus, numerous studies are based on imaging sensory and interneurons in response to mechanical, electrical, chemical, or thermal stimuli (Biron et al., 2008; Gabel et al., 2007; Hilliard et al., 2005; Li et al., 2006; Suzuki et al., 2003). These approaches have the advantage of experimental control over the input so that poststimulus calcium responses can be collected and averaged in a precise manner. Full understanding of how a nervous system encodes information about its environment requires this fine resolution approach, particularly in small nervous systems where one or two neurons may subserve an entire sensory apparatus for the organism. For example, it is well known from behavioral studies that C. elegans is a thermosensitive organism and perhaps as a mechanism of thermoregulation while navigating in soil, animals move down temperature gradients toward the temperature to which they have been adapted (cryotaxis), and navigate along preferred isotherms within a temperature gradient (isothermal tracking) (Clark et al., 2007a; Hedgecock and Russell, 1975; Mori et al., 2007; Ramot et al., 2008b). Genetics, laser ablation, and electrophysiological studies have identified the ciliated sensory neuron AFD as a primary thermosensory neuron and its sole synaptic output is directed onto the interneuron AIY (Mori et al., 2007; Ramot et al., 2008a; White et al., 1986). Calcium imaging in AFD in response to subtle changes in ambient temperature have revealed that this modality is exquisitely sensitive with AFD intracellular calcium continuously tracking temperature shifts as small as 0.05 °C (Clark et al., 2007b). In addition to AFD, the olfactory sensory neuron AWC also contributes to thermosensation based on calcium imaging in response to temperature shifts (Biron et al., 2008; Kuhara et al., 2008). AWC and AFD appear to have distinct responses to temperature changes suggesting that they encode separate aspects of temperature dynamics, and their resulting activity shapes the worm's thermotactic behavior.

Sensory neurons in C. *elegans* are formed as bilaterally symmetric pairs, but this does not mean that they have identical functions. For example, GECIs have been useful in characterizing a functional asymmetry between a pair of gustatory neurons responsible for chemoattraction to NaCl (Suzuki et al., 2008). ASEL and ASER are the left and right homologous pair of chemosensory neurons that detect ions such as sodium and chloride and drive a chemoattractive behavior in the worm (Pierce-Shimomura et al., 2001). Increases in sodium concentration evoked calcium transients in ASEL while decreases in chloride concentration evoked calcium transients in ASER (Fig. 2.2), reminiscent of ON and OFF cells in the vertebrate visual system (Schiller, 1992; Suzuki et al., 2008). This computational strategy is recapitulated in the interneurons AIB and AIY which are downstream of the olfactory neuron pair AWC. AWC and AIB neurons displayed calcium transients upon removal of the stimulus odorant (Fig. 2.3) while AIY calcium transients were evoked by odorant addition (Chalasani et al., 2007). Interestingly, the calcium transients differ in all three neuronal classes with AWC responding transiently to odorant removal, AIB sustaining a prolonged odor-OFF response, and AIY showing a stochastic odor-ON response in its neurite rather than cell body (Chalasani et al., 2007). Thus, GECIs provide spatial and temporal information about the fine structure of olfactory encoding that were previously unavailable. Other examples of studies employing GECIs in the worm nervous system include work on mechanosensation in ALM, proprioception in DVA, and nociception in ASH as well as pioneering studies of calcium dynamics in the pharynx and in the circuitry underlying egg laying (Hilliard et al., 2005; Kerr, 2006; Kerr et al., 2000; Li et al., 2006; Suzuki et al., 2003; Zhang et al., 2008).

C. Limitations of GECIs

As with more conventional calcium indicators, there are a few major limitations to the performance of GECIs as surrogates for intracellular calcium. First, the kinetics of calcium association and dissociation limit the temporal scale on which calcium transients can be measured. Rapid changes in calcium concentration (local presynaptic calcium that triggers transmitter release for instance) will not be reported accurately given that the calcium dissociation kinetics of GECIs are on the order of hundreds of milliseconds. Second, the limited dynamic range of typical GECIs may not allow for a detectable signal especially in the small processes of worm neurons where the absolute number of GECI proteins may be quite small (See Section X). G-CaMP variants generally have a broader dynamic range than cameleons (Kerr and Schafer, 2006; Pologruto et al., 2004;

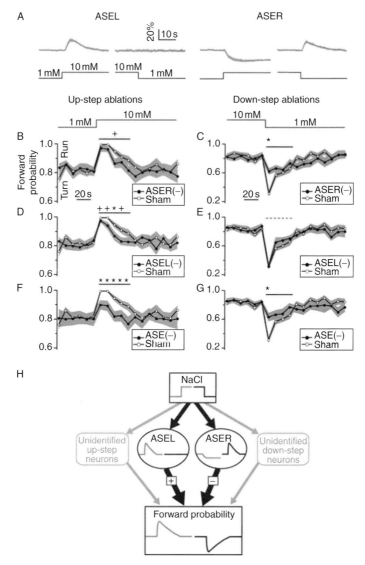

Figure 2.2. Roles of ASEL and ASER in NaCl step-response behavior. (A) Average ASE calcium transients in response to concentration steps of 69 mM. The concentration step is indicated below the first calcium trace. The gray band represents ±1 SEM; $n \geq 5$ recordings, with one recording per worm. (B–G) Effects of unilateral and bilateral ASE ablations on the behavioral response to concentration steps. The probability of forward locomotion is plotted against the time relative to the step. Statistical significance (ablation vs. sham operated) was assessed by means of a repeated measures ANOVA over the indicated time window (horizontal lines above the traces) after the

Reiff *et al.*, 2005). Third, fluorescence can be sensitive to other ions such as protons, chloride, or magnesium thereby confounding changes in intracellular calcium. Current GECIs have been greatly improved in this category. Finally, high expression of GECIs may buffer intracellular calcium and perturb calmodulin-dependent signaling in the host cells. At the same time, low expression may thwart accurate fluorometric measures altogether. Despite these limitations, GECIs have already had an enormous impact on our ability to interrogate the nematode nervous system and will undoubtedly be essential tools for future studies.

IV. IMAGING PRESYNAPTIC TERMINALS

Presynaptic terminals are axonal specializations that release neurotransmitter onto their postsynaptic targets. These subcellular compartments are typically small (about 1 μm) and packed with transmitter-containing SVs as well as hundreds of specialized trafficking proteins responsible for regulating neurotransmission. One common method of visualization employs immunofluorescence imaging after staining with an antibody raised against a presynaptic protein such as synapsin or synaptobrevin. With the advent of GFP-fusion proteins, expression of a GFP-tagged presynaptic protein can provide a convenient means of marking presynaptic terminals. In *C. elegans*, this approach provides an excellent method for marking presynaptic boutons within the ventral and dorsal nerve cords, where *en passant* synapses are sufficiently spread out to resolve individual boutons (Dittman and Kaplan, 2006; Jorgensen *et al.*, 1995; Nonet, 1999; Sieburth *et al.*, 2005; White *et al.*, 1976).

A. Quantifying presynaptic proteins

As our knowledge of the synaptic proteome has grown over the past decade, the repertoire of proteins available as presynaptic markers has expanded considerably (Ch'ng *et al.*, 2008; Jahn *et al.*, 2003; Patel *et al.*, 2006; Weimer and

step (shown above B and C). ASEL(–), ASEL ablation; ASER(–), ASER ablation; ASE (–), bilateral ASE ablation; solid horizontal line, ANOVA significant at $P < 0.05$ or less; dashed horizontal line, not significant; asterisks, time points at which there was a significant difference between means after correcting for multiple comparisons (t-test; $P < 0.05$); plus symbols, time points at which there were significant differences in uncorrected t-tests ($P = 0.05$). Imaging and behavioral data are from different individuals. The gray band represents ± 1 SEM, with $n \geq 15$ in each panel. (H) Functional connectivity implied by (B–G) together with imaging data and unilateral activation experiments. Unidentified neurons (shown in gray) account for residual behavior when ASE is ablated. Reproduced from Suzuki *et al.* (2008).

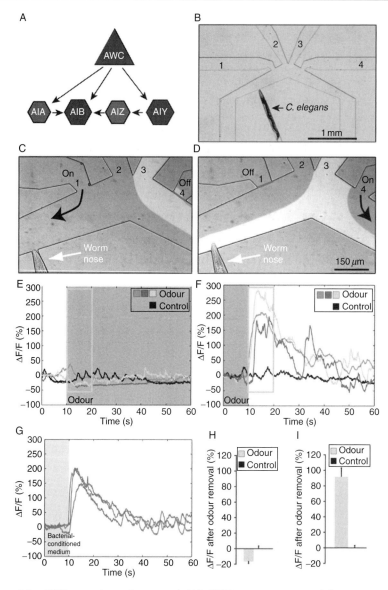

Figure 2.3. AWC responds to odor removal. (A) AWC sensory neurons and downstream interneurons. (B) Low-magnification and (C and D) high-magnification view of the PDMS
imaging chip, with worm nose exposed to buffer (C, streams 1–3 are open; stream
2 reaches the nose) or odor (D, streams 2–4 are open; stream 3 reaches the nose).
(E–G) Representative G-CaMP responses from AWC^{ON}. (E and F) AWC^{ON} responses
on addition (E) or removal (F) of isoamyl alcohol odor (colored traces) or buffer (black

Jorgensen, 2003). When imaging the nerve cords of animals expressing presynaptic GFP-fusion proteins, the amount of fluorescence can be quantified thereby providing information both on the location of a synapse and on how much of a particular presynaptic protein is present. Under ideal circumstances, fluorescence will be proportional to protein concentration in contrast to the nonlinear relationship characteristic of polyvalent antibodies. A dysfunctional synapse may show signs of its impairment in the inappropriate accumulation or loss of certain presynaptic proteins. For instance, in mutant animals where SV exocytosis is impaired, SV markers such as synaptobrevin tend to accumulate at synapses (Dittman and Kaplan, 2006; McEwen *et al.*, 2006; Sieburth *et al.*, 2005). Likewise, these markers become more delocalized in endocytic mutants where SV membrane components accumulate in the axonal plasma membrane (Ch'ng *et al.*, 2008; Dittman and Kaplan, 2006; Jorgensen *et al.*, 1995; Marza *et al.*, 2008; Schuske *et al.*, 2003). Imaging a variety of presynaptic proteins in a particular mutant can shed light on the underlying molecular lesion (Ch'ng *et al.*, 2008; Nonet *et al.*, 1999). When crucial assembly proteins such as the kinase SAD-1 or the PHR protein RPM-1 are removed, synaptic architecture is altered and many presynaptic proteins display altered abundance and spatial distribution (Abrams *et al.*, 2008; Ackley and Jin, 2004; Ch'ng *et al.*, 2008; Crump *et al.*, 2001; Jin, 2005; Margeta *et al.*, 2008; Patel *et al.*, 2006; Zhen *et al.*, 2000). Thus, fluorescently tagged synaptic proteins have proved useful both as indicators of synaptic dysfunction and of disrupted assembly.

B. pH-Sensitive GFP at the synapse

GFP is intrinsically pH sensitive with a single protonation site that modifies its absorption spectrum (Miesenbock *et al.*, 1998; Sankaranarayanan *et al.*, 2000). The pKa of this site is about 6 so GFP that is trafficked to intracellular compartments such as endosomes, lysosomes, and secretory vesicles will be at least partially protonated and its fluorescence quenched; this is the major explanation for the dim fluorescence of GFP-tagged constructs expressed in acidic compartments. Miesenbock and colleagues took advantage of this pH-dependent fluorescence, together with a random mutagenesis strategy to create a GFP derivative with enhanced pH sensitivity and a more neutral pKa of 7.1. This "pHluorin" variant of GFP has become a useful indicator of SV exo- and endocytosis because

---

traces) at 10 s. Gray shading denotes presence of odor; yellow intervals are analyzed in (H) and (I). (G) AWC[ON] responses on removal of bacterial-conditioned medium at 10 s. (H and I) Average fluorescence change in AWC[ON] during the 10 s after odor addition (H) or removal (I) ($n = 30$). Error bars indicate SEM. Reproduced from Chalasani *et al.* (2007). (See Page 3 in Color Plate Section at the end of the book.)

of the pH shift experienced by the lumenal domain of a SV protein (Sankaranarayanan and Ryan, 2000; Sankaranarayanan et al., 2000). pHluorin fused to the C terminus of the vSNARE synaptobrevin (SynaptopHluorin) is about 20-fold brighter on the plasma membrane than inside an acidic SV so the spatial redistribution of SynaptopHluorin that occurs during the SV cycle can be detected by monitoring fluorescence intensity even though this redistribution occurs on a scale well below the spatial limit of resolution for light microscopy. Multiple studies in C. elegans and in rodent hippocampal cultures have reported a substantial pool of synaptobrevin on the plasma membrane (15–30% of total axonal synaptobrevin) that exchanges with vesicular synaptobrevin during ongoing synaptic activity (Dittman and Kaplan, 2006; Fernandez-Alfonso et al., 2006; Li and Murthy, 2001; Wienisch and Klingauf, 2006). This plasma membrane pool of SynaptopHluorin accounts for the majority of fluorescence detected in motor neuron axons of the dorsal cord in C. elegans (Dittman and Kaplan, 2006). Impairment of vesicle exocytosis depletes this pool whereas disruption of endocytosis causes an accumulation of surface SynaptopHluorin and a corresponding increase in fluorescence. SynaptopHluorin fluorescence therefore provides a measure of the balance between exo- and endocytosis in dorsal cord motor neurons. This technique has also been used to follow synaptic activity of the thermosensory AFD neuron and its response to temperature shifts (Samuel et al., 2003).

There are two methodological issues to consider when using pH-sensitive GFPs as indicators of synaptic activity. First, comparisons of absolute fluorescence between wild-type and mutant animals must take into account changes in transgene abundance since mutations that alter synaptic activity may also affect the turnover of SNARE proteins. Relative surface abundance of SynaptopHluorin independent of total protein levels can be estimated in dissected worm preparations by comparing quenched fluorescence in an acidic buffer with unquenched fluorescence in ammonium chloride (Dittman and Kaplan, 2006). For intact animals, one can image N-terminally tagged (cytoplasmic domain) synaptobrevin to monitor changes in protein abundance since the fluorescence will be independent of vesicular or plasma membrane localization. In theory, synaptobrevin could be doubly tagged with N-terminal RFP and C-terminal pHluorin so that the ratio of green to red signal would be proportional to surface abundance. This approach assumes that the doubly tagged vSNARE would be sorted correctly and trafficked to SVs. Regardless of the technique used, another signal contaminant that may confound pHluorin measurements is autofluorescence. C. elegans tissue is autofluorescent in the GFP emission spectrum and this background fluorescence is partially quenched at acidic pH, further complicating the separation of pHluorin signal from background (Dittman and Kaplan, 2006).

A second important consideration arises from the observation that SynaptopHluorin and other pHluorin-fusion proteins are not exclusively localized to presynaptic terminals. Dendritic compartments will also accumulate some SynaptopHluorin so imaging in a region that has intermixed axons and dendrites expressing the transgene will diminish the correlation between fluorescence intensity and the SV cycle. For example, expression of SynaptopHluorin in cholinergic motor neurons will result in a pure axonal signal in the dorsal cord whereas the ventral cord will contain a combination of axonal and dendritic sources. However, GABAergic motor neuron expression of SynaptopHluorin will result in a mixed signal in both the dorsal and ventral cords. Thus, it is critical to consider the cellular anatomy being imaged when interpreting SynaptopHluorin fluorescence.

C. Spatial localization as a signal

The spatial localization and subcellular distribution of a protein may be useful in the optical detection of neuronal activity. For instance, GFP fused to the PH domain of PLCδ1 has been used as a detector of PIP_2 on the inner leaflet of the plasma membrane through its relatively specific binding (Holz and Axelrod, 2002; Micheva *et al.*, 2001). Redistribution of PH-GFP into the cytoplasm occurs upon hydrolysis of PIP_2 into DAG and IP3 by phospholipase Cβ. Local production of DAG and IP3 has been monitored by following the ratio of membrane to cytoplasmic PH-GFP fluorescence. In *C. elegans*, localization of the synaptic protein UNC-13 has been used to follow modulatory changes in synaptic activity resulting from DAG and Gαo signaling. UNC-13S::YFP became more enriched at presynaptic terminals when either the DAG Kinase DGK-1 or the Gαo subunit GOA-1 were removed, and this recruitment was dependent on the DAG-binding C1 domain in UNC-13 (Lackner *et al.*, 1999; Nurrish *et al.*, 1999). Inhibitory modulators acting through the Gαo pathway caused a dispersion of synaptic UNC-13S::YFP suggesting a possible molecular mechanism for some of the inhibitory effects of modulators such as serotonin. Removal of the SNARE negative regulator Tomosyn also resulted in synaptic recruitment of UNC-13S::YFP (McEwen *et al.*, 2006). This enrichment of UNC-13 may account for some of the enhanced transmitter release observed in Tomosyn *tom-1* mutants (Dybbs *et al.*, 2005; Gracheva *et al.*, 2006; McEwen *et al.*, 2006). These examples highlight the capacity of spatial localization to convey important information on the state of a neuron or individual presynaptic terminal in terms of important modulatory pathways such as DAG signaling.

D. Systematic profiling of presynaptic proteins

Given that the abundance and localization of presynaptic proteins provides information on the state of the synapse, proteins with similar functions or shared binding partners may have correlated changes in their synaptic distributions. This approach was used by Sieburth *et al.* so provide an initial characterization of novel synaptic proteins found in an RNA interference screen (Sieburth et al., 2005). One can imagine further systematizing this approach so that the correlation across a panel of tagged synaptic proteins can be computed for a given synaptic mutation. Ch'ng et al. utilized this approach with nine GFP-tagged synaptic proteins imaged in 25 mutants that affect synaptic transmission (Ch'ng et al., 2008). By quantifying the distribution and abundance of these nine probes in each of the mutants, the authors were able to cluster genes based on similar imaging profiles (Fig. 2.4). For instance, genes required for exocytosis and genes responsible for synapse formation clustered into distinct groups. Furthermore, genes that fell into a cluster corresponding to dense core vesicle (DCV) secretion also regulated animal lifespan in an insulin-dependent manner. Thus, there is likely to be information contained not only in quantification of individual optical probes but also in correlations between distinct optical probes.

V. IMAGING POSTSYNAPTIC COMPARTMENTS

Regulation of transmitter release is only one aspect of synaptic transmission and its plasticity. The postsynaptic compartment is responsible for transforming the extracellular chemical signal (neurotransmitter) into electrical and chemical signals in the neuron or muscle receiving the synaptic input. Multiple aspects of this postsynaptic compartment play a role in the regulation of synaptic strength. For instance, synaptic receptors directly determine the magnitude and time course of current flow across the plasma membrane. Thus, changes in receptor number or conductance will alter the impact of transmitter release. In addition to the receptor current, voltage-gated ion channels and second messenger systems triggered by G-protein coupled receptors will shape the postsynaptic response. Calcium influx initiates enzymatic cascades and transcriptional signaling networks, altering neuronal activity on longer time scales. In the mammalian brain, regulation of postsynaptic receptors is an important means of expressing long-term changes in synaptic strength that are thought to underlie memory storage (Malinow, 2003; Malinow and Malenka, 2002).

A. Postsynaptic glutamate receptors

Glutamate is a major excitatory neurotransmitter in C. elegans as well as in vertebrates, and there are several classes of postsynaptic receptors that mediate its synaptic signaling. These can be generally divided into metabotropic glutamate receptors (mGluRs) and ionotropic glutamate receptors (iGluRs). In nematodes, glutamate also activates a chloride conductance (GluCl channel). There are eight known non-NMDA class iGluRs (GLR-1–GLR-8) and two NMDA class iGluRs (NMR-1 and NMR-2) expressed in multiple neurons in

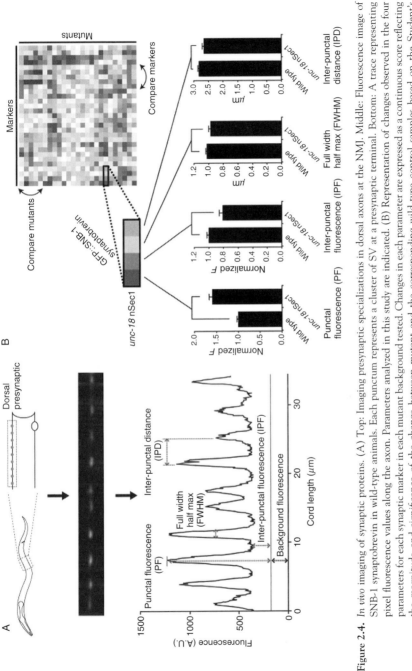

Figure 2.4. *In vivo* imaging of synaptic proteins. (A) Top: Imaging presynaptic specializations in dorsal axons at the NMJ. Middle: Fluorescence image of SNB-1 synaptobrevin in wild-type animals. Each punctum represents a cluster of SV at a presynaptic terminal. Bottom: A trace representing pixel fluorescence values along the axon. Parameters analyzed in this study are indicated. (B) Representation of changes observed in the four parameters for each synaptic marker in each mutant background tested. Changes in each parameter are expressed as a continuous score reflecting the magnitude and significance of the change between mutant and the corresponding wild-type control samples based on the Student's *T*-statistic. Positive scores (red shading) and negative scores (blue shading) indicate an increase or decrease, respectively, in a given parameter in the mutant compared to wild type. The magnitude of the score is indicated by the intensity of the shading. How *unc-18* nSec1 mutants affected SNB-1 synaptobrevin is used as an example. Error bars are SEM. Reproduced from Ch'ng *et al.* (2008). (See Page 4 in Color Plate Section at the end of the book.)

the worm nervous system (Brockie and Maricq, 2006). The GLR-1 subunit can be fused to GFP at its C terminus prior to the group I PDZ ligand motif TAV (Rongo *et al.*, 1998). This GLR-1 GFP-fusion protein is functional and localizes to synapses, thereby providing a fluorescent probe for the postsynaptic compartment in worm interneurons (Burbea *et al.*, 2002; Rongo *et al.*, 1998). Monitoring the receptor abundance at each synapse as well as the degree of localization and synaptic density allows for a sensitive quantification of the glutamate receptor pool (Fig. 2.5). This approach has been useful for investigating the regulation of glutamate receptor abundance by AP180, ubiquitin, and multiple E3 ligases (Burbea *et al.*, 2002; Dreier *et al.*, 2005; Juo and Kaplan, 2004; Schaefer and Rongo, 2006). Glutamate receptor dynamics have been described in other systems as well, using a variety of optical techniques and tagged receptors

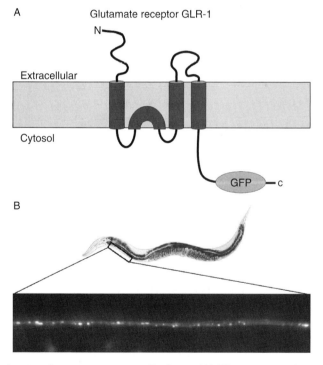

Figure 2.5. Imaging glutamate receptors in *C. elegans.* (A) The ionotropic glutamate receptor GLR-1 is fused with GFP near its C terminus. Note that the C-terminal TAV occurs after insertion of the GFP. (B) Fluorescence image taken from the ventral nerve cord near the animal's head. Note the punctate distribution of GLR-1::GFP fluorescence. Scale bar is 10 μm. Figure contributed by Lars Dreier, Department of Neurobiology, UCLA.

(Ashby *et al.*, 2004; Cognet *et al.*, 2006; Lin and Huganir, 2007). Given the success of fluorescently tagged receptors in C. *elegans* and other systems, this approach will likely continue to provide insights into the postsynaptic compartment and the cell biology that underlies its regulation.

B. Postsynaptic nicotinic acetylcholine receptors

Another class of ionotropic receptor that has been imaged at the synapse in C. *elegans* is the nicotinic acetylcholine receptor (nAChR). This class of neurotransmitter receptor mediates synaptic transmission at NMJs, and is therefore extensively expressed throughout the motor nervous system. The worm genome contains a large gene family for nAChRs with at least 27 subunit genes (Jones and Sattelle, 2004). Canonical nAChRs are assembled as either homo- or heteropentamers containing ACh-binding alpha subunits along with nonalpha subunits (Rand, 2007). At body wall muscle NMJs, nAChRs are composed of two pharmacological classes: the nicotine-sensitive ACR-16 homomultimers and the levamisole-sensitive LEV-1:UNC-29:UNC-38/UNC-63 heteromultimers (Boulin *et al.*, 2008; Rand, 2007; Richmond and Jorgensen, 1999; Touroutine *et al.*, 2005). Electrophysiological recordings at the NMJ have been invaluable in elucidating the basic properties of a C. *elegans* synapse and revealing the profound conservation between nematode and vertebrate synaptic biology (Gracheva *et al.*, 2006; Madison *et al.*, 2005; McEwen *et al.*, 2006; Richmond and Jorgensen, 1999; Richmond *et al.*, 1999; Wang *et al.*, 2001; Weimer *et al.*, 2003). These recordings provide a direct window into the abundance and properties of surface synaptic nAChRs. Imaging tagged receptors adds to our understanding of the postsynaptic compartment by revealing information on the entire receptor pool, including endosomal, extrasynaptic, and inactive receptors. A variety of signaling pathways in the muscle have been found to regulate receptor abundance and ultimately control the locomotory performance of the worm. For example, muscle receptors have been visualized *in situ* by both GFP tagging and by injecting antibodies into the worm pseudocoelom (Francis *et al.*, 2005; Gally *et al.*, 2004; Gottschalk and Schafer, 2006; Gottschalk *et al.*, 2005). Levamisole receptor abundance is regulated by the receptor tyrosine kinases SOC-1 and CAM-1 and synaptic receptor clustering is controlled by the CUB/LDL transmembrane protein LEV-10 (Gally *et al.*, 2004; Gottschalk *et al.*, 2005). Synaptic ACR-16 is also regulated by CAM-1 (Francis *et al.*, 2005). The microRNA miR-1 regulates the levamisole receptor possibly by altering its subunit composition (Simon *et al.*, 2008). Thus, for questions about trafficking and recycling of receptors or about synapse assembly and disassembly, having access to the entire pool of receptors is crucial.

C. Other receptors

Although much attention has been paid to the ionotropic glutamate and ACh receptors summarized above, C. elegans depends on a host of other receptor types for proper nervous system function. Bamber et al. investigated the composition of GABA$_A$ receptors at the NMJ through a combination of electrophysiology and imaging of receptor subunit localization (Bamber et al., 2005). The localization and trafficking of GABA$_A$R-GFP was found to depend on GABAergic and cholinergic inputs onto muscle, and failure to form synapses onto a muscle cell resulted in GABA$_A$R-GFP trafficking to autophagosomes (Rowland et al., 2006). Metabotropic receptors also play an important role in the worm nervous system but almost nothing is known about their subcellular localization in most cases. The G-protein coupled ACh receptor GAR-2 tagged with YFP was imaged in worm neurons and found to be diffusely expressed throughout motor neuron axons in the dorsal cord instead of being restricted to cholinergic synapses (Dittman and Kaplan, 2008). Perhaps this distribution reflects a role in detecting low concentrations of extrasynaptic ACh. Although receptors indicate the site of signal transduction, we know very little about the spatial spread or restriction of downstream signaling. Future studies using optical probes based on signaling molecules will aid in improving our understanding of information processing in the worm nervous system.

VI. IMAGING PROXIES FOR NEUROPEPTIDE SECRETION

One important aspect of neurosecretory function which has not received the same attention as synaptic transmission is neuropeptide secretion from DCVs. The C. elegans genome contains 113 neuropeptide genes and neuropeptide mutants have a variety of behavioral defects including locomotion, egg laying, and dauer formation (Li and Kim, 2008; Li et al., 1999). Despite the importance of neuropeptide release in worms as well as other animals, less is known about the process and its regulation because DCV fusion and cargo release is difficult to measure in many experimental systems compared to conventional chemical transmission. DCVs do not undergo synchronous fusion in large numbers upon stimulation in neurons in contrast to small SVs. In addition, neuropeptides typically act in a hormonal fashion, far from the site of release as opposed to small neurotransmitter molecules which directly gate ion channels in the postsynaptic membrane. One approach that has proved useful for monitoring DCVs and neuropeptide release has been to express neuropeptides tagged with fluorescent proteins such as YFP Venus. This fluorescent cargo is secreted into the pseudocoelom where it can diffuse to distant targets (Fig. 2.6). Scavenger cells known as coelomocytes reside in this space and continuously endocytose

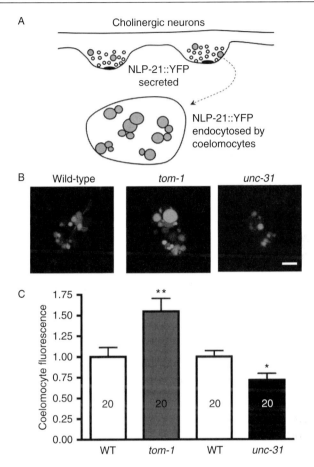

Figure 2.6. *tom-1* and *unc-31* mutants exhibit altered peptide release. (A) Schematic of the peptide secretion assay, in which YFP(Venus)-tagged neuropeptide NLP-21 (KP#1383 *punc-129::nlp-21::Venus*), once released from cholinergic neuron DCVs, is taken up by scavenger cells known as coelomocytes. (B) Representative images of a single coelomocyte from wild-type, *tom-1(ok285)*, and *unc-31(e928)* young adult worms showing multiple lysosomes containing NLP-21::YFP previously secreted from cholinergic neurons. Scale bar, 10μm. C, Average fluorescent intensity values (normalized to wild type (WT) matched for image acquisition parameters) measured from the posterior ventral coelomocyte of wild type, *tom-1(ok285)*, and *unc-31(e928)* worms expressing NLP-21:YFP in cholinergic neurons. The significance values (asterisks) above the bars are relative to the wild type, and numbers in bars represent the sample size (*n*). Reproduced from Gracheva *et al.* (2007). (See Page 5 in Color Plate Section at the end of the book.)

pseudocoelomic fluid; therefore take up fluorescent peptides as well (Fares and Greenwald, 2001a,b; Zhang *et al.*, 2001). Multiple labs have taken advantage of this scavenger activity to follow steady-state neuropeptide release as measured by fluorescence accumulation in coelomocytes (Gracheva *et al.*, 2007; Sieburth *et al.*, 2007; Speese *et al.*, 2007). By imaging fluorescence in axons and coelomocytes, it has been possible to demonstrate that neuropeptides like NLP-21 normally collect in coelomocytes. Disruption of DCV fusion by elimination of an important cytoplasmic molecule involved in DCV fusion (UNC-31 CAPS) causes increased axonal fluorescence of NLP-21 Venus with concomitant decrease in coelomocyte fluorescence. Increasing DCV fusion by removal of a negative regulator (tomosyn) resulted in less axonal fluorescence and a reciprocal increase in coelomocyte fluorescence (Ch'ng *et al.*, 2008; Gracheva *et al.*, 2007).

VII. BIMOLECULAR FLUORESCENCE COMPLEMENTATION IN THE WORM

A functional GFP can be assembled spontaneously from two partial GFP polypeptides when the peptides are brought into close approximation by other protein interactions (Hiatt *et al.*, 2008; Magliery *et al.*, 2005). For instance, antiparallel leucine zippers fused to the partial GFP polypeptides will dimerize in a cell and permit GFP assembly resulting in fluorescence (Magliery *et al.*, 2005). The appearance of fluorescence therefore indicates some underlying protein–protein interaction. Hynes *et al.* reconstituted GFP fluorescence based on assembly of the G-protein beta gamma heterodimer in cell culture (Hynes *et al.*, 2004). Termed bimolecular fluorescence complementation (BiFC), this general strategy for a fluorescent coincidence detector has already found multiple uses in *C. elegans*.

A. Combinatorial promoters

One of the first powerful uses of GFP in *C. elegans* was as a reporter for gene expression (Boulin *et al.*, 2006; Hobert and Loria, 2006). The expression pattern of GFP driven by *cis*-regulatory elements from a particular gene provides a means of deducing where that gene product is normally expressed. Zhang *et al.* extended the use of these reporter constructs by using two distinct promoter regions each driving a partial GFP polypeptide attached to a leucine zipper motif (Zhang *et al.*, 2004). Because fluorescence will only be detected in cells expressing both halves of GFP, one can now identify subsets of cells specified in common by both *cis*-regulatory elements. The authors suggest a combinatorial approach to identifying expression patterns for an unknown gene based on a series of strains expressing a partial GFP, CFP, or YFP polypeptide each under a well-described promoter. If the unknown promoter is used to drive the complementary peptide in a separate strain, then crossing this strain to each of the starting strains will produce a different pattern of fluorescence, leaving less ambiguity as to which cells express the unknown gene (Zhang *et al.*, 2004).

B. GRASP

A recent modification of the BiFC strategy has been used to produce an optical indicator of synaptic connectivity in C. *elegans*. Feinberg *et al.* introduced a labeling technique termed "GFP Reconstitution Across Synaptic Partners" or GRASP (Feinberg *et al.*, 2008). The basic principle of GRASP is similar to the BiFC described above except that each half of GFP is fused to a transmembrane protein on the extracellular face of two separate cells (Fig. 2.7). Since an unfolded partial GFP polypeptide will often be degraded in the endoplasmic reticulum, an asymmetric split GFP system was developed that uses the first 214 residues of a fast-folding GFP variant as one piece (Cabantous *et al.*, 2005). This piece folds well by including 10 of the 11 strands of the GFP beta-barrel. The remaining strand is the complementary fragment of GFP, a peptide fragment consisting of only 16 residues. Expression of these complementary fragments using a generic transmembrane tag (CD4) essentially created a proximity detector whereby processes that came into close contact developed a continuous fluorescent label at the site of contact. However, if a localized presynaptic protein was used as the carrier for one of the fragments, specific synaptic fluorescence could be observed. For example, the authors used a presynaptic transmembrane phosphatase (PTP-3A) tagged with the 11th barrel strand expressed in the HSN neuron together with CD4 fused to the first 10 strands of GFP expressed in either of two HSN target cells: VC motor neurons or vulval muscle (Feinberg *et al.*, 2008). When CD4-GFP(1–10) was expressed in vulval muscle, fluorescence was only observed at a few specific points of contact between HSN and its target muscle. Moreover, these sites were distinct from fluorescent sites observed when CD4-GFP(1–10) was expressed in VC motor neurons, suggesting that these two separate synaptic outputs of HSN occurred at different points along the HSN axon. This synaptic connection pattern was confirmed by examining electron micrographs of HSN synapses, supporting the idea that GRASP fluorescence corresponded to bona fide synaptic connections (Feinberg *et al.*, 2008). GRASP promises to be a useful fluorescence technique for identifying and quantifying synaptic connectivity in the living animal with high resolution.

VIII. METHODS FOR IMMOBILIZING THE WORM

As mentioned previously, the ability to image the nervous system of intact living animals is a distinct advantage of C. *elegans* as an experimental system. This advantage also comes with the experimental challenge of either imaging an animal while it is in motion or immobilizing the animal while possibly abolishing the behavior one is attempting to study.

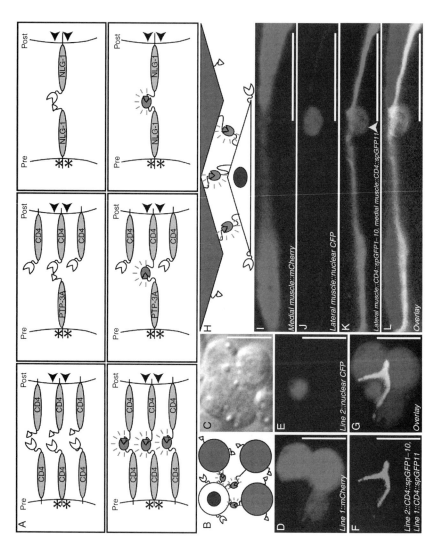

Figure 2.7. Continued

A. Pharmacological paralysis

For high-resolution imaging studies such as measuring presynaptic protein distributions, the chief approach has been to pharmacologically immobilize animals and image them sandwiched between a coverslip and a hydrated agarose pad. Typical paralytic agents include the anesthetic phenoxypropanol (Schuske *et al.*, 2003; Sulston and Horvitz, 1977), the mitochondrial poison azide (Murfitt *et al.*, 1976; Nonet, 1999; Nurrish *et al.*, 1999), the nAChR agonist levamisole/tetramisole (Burbea *et al.*, 2002; Juo and Kaplan, 2004; Lewis *et al.*, 1980a,b), and the myosin blocker 2,3 butanedione monoxime (Dittman and Kaplan, 2006; Sieburth *et al.*, 2005). Worms immobilized with these agents can be imaged for brief periods (10–30 min) while their neurons remain relatively healthy. However, for prolonged time-lapse imaging studies, viability becomes a major issue. Furthermore, few if any behaviors will be expressed in this state of paralysis so imaging circuit-level activity will not be possible.

B. Glue

An alternative approach that has been used for calcium imaging while mechanically stimulating the worm is to glue the animal onto a coverslip directly, usually with cyanoacrylate glue. This has the advantage of allowing a functional nervous system to respond to an applied stimulus (Faumont and Lockery, 2006; Li *et al.*, 2006; Suzuki *et al.*, 2003; Zhang *et al.*, 2008). However, it is difficult to use on a large scale as each animal must be carefully glued down by hand, and recovery of the animals is seldom successful. In addition, some behaviors such as locomotion will be profoundly affected by gluing the cuticle to a surface.

Figure 2.7. GRASP strategy and demonstration of extracellular GFP assembly *in vitro* and *in vivo* (A) Schematic diagram of GRASP with delocalized CD4 tethers (left), presynaptically localized PTP-3A and a delocalized CD4 tether (center), and pre- and postsynaptically localized NLG-1 tethers (right). Asterisk symbolizes presynaptic site; arrowhead, postsynaptic site. (B–G) Extracellular GFP reconstitution in culture. Three cells express mCherry and CD4::spGFP11, and one cell expresses nuclear CFP and CD4::spGFP1–10. (B) Schematic diagram. (C) Differential interference contrast image. (D) mCherry. (E) Nuclear CFP. (F) GRASP GFP signal. (G) Merge. Body wall muscle cells were labeled using the *myo-3* promoter. (H–L) Extracellular GFP reconstitution *in vivo*. (H) Schematic drawing of two rows of dorsal body wall muscles. Medial muscle cells express mCherry and CD4::spGFP11, and lateral muscle cells express nuclear CFP and CD4:: spGFP1–10. (I) mCherry. (J) Nuclear CFP. (K) GRASP GFP signal; yellow arrowhead marks CFP bleed-through. (L) Merge. Medial dorsal body wall muscle cells were labeled using the *ace-4* promoter, and lateral body wall muscle cells using the *him-4* promoter. Scale bars are 5 μm in (C)–(G), 10 μm in (I)–(L). Reproduced from Feinberg *et al.* (2008). (See Page 6 in Color Plate Section at the end of the book.)

C. Microfluidics

In the past few years, multiple groups have developed microfluidic devices for reversible immobilization of and environmental control over individual animals (Chronis et al., 2007; Chung et al., 2008; Guo et al., 2008; Hulme et al., 2007). Worms can be trapped in small PDMS chambers microfabricated by soft lithography. While immobilized, worms can then be imaged with high-spatial resolution in the absence of glue or drugs. Some groups have emphasized designs for precise delivery of sensory stimuli while imaging the responses of sensory and interneurons (Chalasani et al., 2007; Chronis et al., 2007) or performing nano-surgery on neuronal processes (Guo et al., 2008), while others devised chambers for trapping a large number of animals in parallel for systematic imaging and sorting (Chung et al., 2008; Hulme et al., 2007; Rohde et al., 2007). Cui et al. combined PDMS microfluidics with a CMOS light sensor array to create a lensless, on-chip microscope for high-resolution imaging of C. elegans (Cui et al., 2008). The fabrication of structured microenvironments in which the worm navigates a defined obstacle course now allows for high-resolution imaging of a freely moving animal (Lockery et al., 2008; Park et al., 2008). Given the rapid proliferation of microfluidic devices dedicated to the handling of C. elegans, this approach is likely to become a standard part of the worm neurobiologist's toolkit.

D. Imaging on the go

Clark et al. performed calcium-imaging experiments on freely moving worms utilizing an improved variant of cameleon (YC 3.60) which has a nearly sixfold signal enhancement over YC 2.12 (Clark et al., 2007a). In this experiment, the thermosensory neuron AFD was imaged in unrestrained animals placed in a temperature gradient with an X–Y motorized stage controlled by the experimenter. Using only a 0.75 NA objective (20× magnification) and a conventional CCD camera, high-quality calcium measurements could be collected while animals freely executed thermotactic behaviors (see Section III.B). The AFD calcium signal was found to carry sufficient information on whether the worm was heading up or down a temperature gradient, and therefore in principle AFD could activate the appropriate motor responses based on its activity (Fig. 2.8). Imaging freely moving animals will be important for understanding precisely how sensory information is collected and processed by the nervous system as a whole. This is particularly relevant when sensory processing depends explicitly on the active participation of motor function during the act of sensation. Immobilization and paralysis will most likely alter the role of motor feedback during sensory processing.

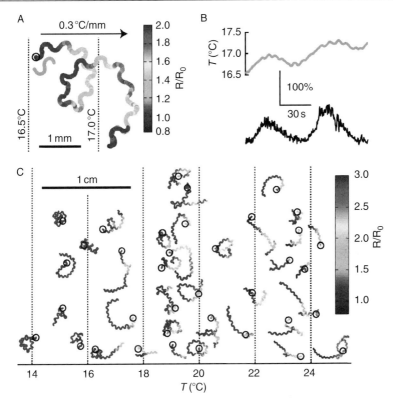

Figure 2.8. AFD neuronal activity in unrestrained worms navigating spatial thermal gradients. (A) The undulating path of an unrestrained worm crawling ~8 mm on a spatial temperature gradient in the vicinity of T^*_{AFD}. The color and position of each data point represent the ratiometric emission signal and position of the AFD thermosensory neuron of the moving worm, respectively. The black circle indicates initial position. To indicate scale, the gray silhouette shows the approximate relative size of the worm body. (B) The top and bottom panels show the time-varying temperature at the worm's head and the ratiometric emission signal from AFD for the data shown in (A). (C) A compilation of data showing the ratiometric emission signals for worms cultivated at 20 °C navigating spatial thermal gradients at different absolute temperatures from ~15 to ~25 °C. The ratiometric emission signal exhibits position dependence only at temperatures above ~17 °C (i.e., $T > T^*_{AFD}$). A black circle indicates the initial position of each data trace. Reproduced from Clark *et al.* (2007a). (See Page 7 in Color Plate Section at the end of the book.)

IX. FEMTOSECOND LASERS AND NANOSURGERY ON THE WORM

Given the known wiring diagram of the worm nervous system, it is tempting to perform a systematic dissection by severing particular neuronal processes and assaying the behavioral consequences. Questions regarding the structure and robustness of the nematode neural network as well as questions about the ability to regenerate connections in the adult animal could be addressed in such a manner. Although in principle, one could use a strategy similar to laser ablation and focus light onto a neurite instead of a cell body, in practice the processes of many neurons are fasciculated together too closely for conventional laser ablation. Recently, an interesting variation on this theme emerged when femtosecond-pulsed lasers were trained onto submicron diameter processes in C. elegans (Chung et al., 2006; Yanik et al., 2004). Because of the extremely short-lived, high-intensity nature of illumination with these pulsed lasers, photodamage could be reduced to the submicron level. Creation of a brief nanoscale plasma and ensuing cavitation bubble (less than 20 aL in volume, 300 nm in diameter) inside a single neuronal process was found to sever the process after a 2-min train of kilohertz-pulsed infrared illumination at 3 nJ with no collateral damage (Chung et al., 2006). Ultraviolet nanosecond pulses have also been shown to sever worm neurites in vivo (Rao et al., 2008). Thus, the capability to selectively disconnect a particular neuronal connection or region of a neuron has now been realized.

A. Beyond cell ablation

One application of femtosecond-pulsed laser ablation (FPLA) is the severing of sensory dendrites from their cell bodies. In a study on the thermosensory properties of AFD, it was found the afferent processes of AFD neurons retained their ability to track small temperature changes after being severed from the AFD cell body (Clark et al., 2006). Furthermore, AFD neurons exhibit a form of plasticity where the cultivation temperature sets the threshold for sensitivity to small temperature shifts (Kuhara et al., 2008; Mori et al., 2007; Ryu and Samuel, 2002). The AFD dendritic process retained this temperature set point based on calcium imaging after being severed from the cell body, indicating that both the thermotransduction and its plasticity are local properties of the sensory process rather than originating from the soma or from other synaptic inputs (Clark et al., 2006). A similar approach was taken with the AWC neuron in order to demonstrate that the thermosensory activity of AWC depends on its dendritic process (Biron et al., 2008).

B. Axonal outgrowth and repair

Many adult neurons in both vertebrates and invertebrates have the capacity to regenerate processes after injury (Cebria, 2007; Chen et al., 2007). Importantly, mammalian central neurons appear to be inhibited from engaging in this process in part by myelin-dependent signaling pathways (McGee and Strittmatter, 2003). Multiple groups in the C. elegans research community have employed

FPLA to investigate the nematode's capacity to regenerate axons both during development and in adulthood (Gabel *et al.*, 2008; Wu *et al.*, 2007; Yanik *et al.*, 2004). Wu *et al.* observed axonal regeneration within 24 h for a variety of cell types including motor neurons, chemosensory neurons, and mechanosensory neurons (Wu *et al.*, 2007). In the case of the mechanosensory neuron PLM, the accuracy of rewiring to the proper location and the rate of regrowth declined during development following axotomy. This migration error was partially mitigated in *vab-1* Eph receptor tyrosine kinase mutants, suggesting that regenerating PLM axons are sensitive to ephrin signaling. Gabel *et al.* took a similar approach, using FPLA to sever processes from a variety of sensory and motor neurons (Gabel *et al.*, 2008). They found that the cholinergic motor neurons (DA/DB class) and AVM mechanosensory neurons show robust regeneration while axons from ASH and AWC sensory neurons fail to regenerate after adult-stage axotomy (Fig. 2.9). Detailed analysis of axon terminal dynamics revealed that the proximal process sends out exploratory lamellipodial extensions, some of which successfully reach their ventral target region. Moreover, mutations in both the Netrin and Slit signaling pathways appeared to have differential effects on initial wiring versus rewiring postaxotomy. The findings from these labs indicate that *C. elegans* will likely be a fruitful organism for studies on nerve regeneration in addition to its well-established role as a model for nervous system assembly.

X. DESIGNING THE IDEAL OPTICAL PROBE

When considering which properties of a genetically encoded optical probe are critical for its usefulness, the type of activity being measured will determine many important features of the probe as discussed earlier. However, there are a few general properties that an ideal probe should possess regardless of the neuronal signal being measured. To ensure that the optical signal follows the process being studied, the ideal probe should participate in the signaling process at some level. Otherwise, the connection between the probe behavior and the underlying signaling will not be clear. For instance, GFP-fusion proteins that are monitored to report abundance, localization, and mobility should be functional in whichever processes the protein normally participates. In *C. elegans*, this aspect can be verified by determining whether the GFP-tagged protein can rescue a loss of function mutant for that protein. Furthermore, the probe and detection system should be sensitive enough to detect relatively low expression levels since highly overexpressed probes may disrupt the process being studied as well as perturb other unanticipated processes (Motley *et al.*, 2006; Rappoport and Simon, 2008).

Expression and detection become critical issues when imaging from a small structure and when the fusion protein is expressed at very low abundance. For instance, imaging presynaptic terminals in the worm dorsal cord can be

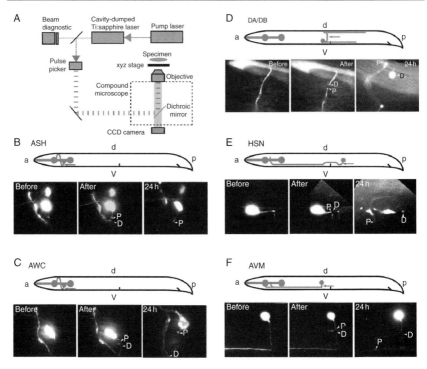

Figure 2.9. Axon regeneration is exhibited by specific neuronal types. (A) Schematic of Femtosecond laser system. (B–F) Axon trajectories of ASH, AWC, DA/DB, HSN, and AVM neurons. In each case, an image is shown before, immediately after and 24 h after surgery. (D) indicates the distal end and P the proximal end of severed axons. Red arrows point to the laser target. ASH and AWC axons were snipped at their posterior ventral projections; ASH axons did not noticeably grow out after 24 h (B; *n* = *12*); AWC axons did not grow out at all (C; *n* = 12). Representative examples of successful axon regeneration in the cholinergic DA/DB motorneurons (D), serotonergic HSN motorneurons (E) and AVM mechanosensory neurons (F). Reproduced from Gabel *et al.* (2008). (See Page 8 in Color Plate Section at the end of the book.)

challenging because a typical bouton has a volume of approximately 0.5 fL and a size approaching the diffraction limit of light microscopy. SNB-1 Synaptobrevin and UNC-10 Rim1, two synaptic markers that have been used at worm synapses provide a quantitative example with GFP as the fluorophore. SNB-1 GFP-fusion proteins provide easily detectable synaptic fluorescence even when very low concentrations of cDNA are used to create the transgenic lines. UNC-10 GFP, however, is much more difficult to detect even when high concentrations of cDNA are injected. Synaptobrevin 1 is normally expressed at about 50–100 copies per SV and there are roughly 100 vesicles in a terminal (Jorgensen *et al.*, 1995; Takamori *et al.*, 2006). This complement of 5000–10,000 SNB-1 molecules in

0.5 fL corresponds to between 15 and 30 μM concentration. Low expression of a GFP Synaptobrevin fusion protein was estimated to be one or two copies per vesicle in one study (Fernandez-Alfonso *et al.*, 2006). This would be the equivalent of 300–600 nM SNB-1::GFP at a bouton. Active zone proteins involved in vesicle fusion are thought to be present at very low copy numbers at the synapse which would be correspond to a few nanomolar (Shapira *et al.*, 2003; Tao-Cheng, 2006). Therefore, even if every UNC-10 Rim1 in a neuron was replaced with an UNC-10 GFP-fusion protein, the fluorophore concentration at the synapse may be no more than a few nanomolar. This spectrum of expression levels provides an approximate dynamic range for detecting fluorescent presynaptic proteins *in vivo* by conventional wide field and scanning confocal microscopy. Fluorophore concentrations below a few nanomolar approach single molecule detection while micromolar concentrations are readily detected. This analysis ignores properties of the fluorophore such as its quantum efficiency, absorption coefficient, or bleach rate. For fluorophores with significantly poorer performance than GFP, detection of a few copies at in a presynaptic terminal may not be possible.

In contrast to the desirable properties of an ideal probe mentioned above, real GFP-fusion proteins are associated with a few well-documented pitfalls that should be considered when designing a candidate probe. Overexpression of a protein can trigger a variety of host-cell responses including the unfolded protein response (UPR), missorting of transmembrane and cargo proteins to inappropriate compartments, and disruption of normal trafficking of bystander proteins. The location of the GFP tag may disrupt protein folding, targeting, or important interactions with other proteins or lipids. Some fluorophores such as wt GFP, dsRed, and tdTomato form multimers possibly causing an unwanted aggregation of the tagged protein. Finally, when quantifying fluorescence relative to a control condition or genotype, it is worth noting that changes in fluorescence are a compound function of the illumination, local fluorophore concentration, synthesis and degradation rates, as well as the cellular geometry. Because alterations in any of these parameters will change the measured fluorescence, quantitative measures of fluorescence should include appropriate controls for such unanticipated changes. For instance, imaging soluble GFP levels under the same promoter can reveal underlying transcriptional changes as well as morphological changes. Normalizing fluorescence to a standard such as a fluorescent slide or metering the light source output will help eliminate artifacts from fluctuations in illumination. Although this list of potential complications may appear to undermine the value of genetically encoded optical probes, the numerous examples of successful and informative studies summarized in this chapter suggest otherwise. In only a few years after their introduction to the C. *elegans* research community, these approaches have revolutionized our understanding of both the nematode nervous system and of neuronal cell biology in general.

XI. CONCLUSIONS AND PROSPECTS

The use of optical probes described in this chapter will advance our understanding of the nervous system on many fronts in the years to come. At the molecular level, use of resonance energy transfer, recovery from photobleaching, and photoconversion will provide detailed information on the microenvironment of proteins and lipids in individual neurons. The ever-growing arsenal of fluorescently tagged proteins will continue to reveal spatial and temporal dynamics of neuronal subcellular structure and signaling. Dynamic probes will delineate the activity and plasticity of neurons across a broad time scale ranging from milliseconds to days. *C. elegans* provides a spectacular arena in which all of these approaches can be developed and perfected in the context of a behaving animal. And, many of the lessons learned in this system, both technical and biological, will undoubtedly transcend the nematode and penetrate deeply into more complex vertebrate nervous systems.

Acknowledgments

The author thanks Josh Kaplan, Queelim Ch'ng, Tim Ryan, Jihong Bai, and Anant Menon for invaluable discussions and critical reading of this chapter, and Lars Dreier for contributing images and unpublished data.

References

Abrams, B., Grill, B., Huang, X., and Jin, Y. (2008). Cellular and molecular determinants targeting the *Caenorhabditis elegans* PHR protein RPM-1 to perisynaptic regions. *Dev. Dyn.* **237**(3), 630–639.

Ackley, B. D., and Jin, Y. (2004). Genetic analysis of synaptic target recognition and assembly. *Trends Neurosci.* **27**(9), 540–547.

Ashby, M. C., Ibaraki, K., and Henley, J. M. (2004). It's green outside: Tracking cell surface proteins with ph-sensitive GFP. *Trends Neurosci.* **27**(5), 257–261.

Bamber, B. A., Richmond, J. E., Otto, J. F., and Jorgensen, E. M. (2005). The composition of the GABA receptor at the *Caenorhabditis elegans* neuromuscular junction. *Br. J. Pharmacol.* **144**(4), 502–509.

Bargmann, C. I. (1998). Neurobiology of the *Caenorhabditis elegans* genome. *Science* **282**(5396), 2028–2033.

Bargmann, C. I., and Horvitz, H. R. (1991). Control of larval development by chemosensory neurons in *Caenorhabditis elegans*. *Science* **251**(4998), 1243–1246.

Bargmann, C. I., and Kaplan, J. M. (1998). Signal transduction in the *Caenorhabditis elegans* nervous system. *Annu. Rev. Neurosci.* **21**, 279–308.

Bargmann, C. I., Hartwieg, E., and Horvitz, H. R. (1993). Odorant-selective genes and neurons mediate olfaction in *C. elegans*. *Cell* **74**(3), 515–527.

Biron, D., Wasserman, S., Thomas, J. H., Samuel, A. D., and Sengupta, P. (2008). An olfactory neuron responds stochastically to temperature and modulates *Caenorhabditis elegans* thermotactic behavior. *Proc. Natl. Acad. Sci. USA* **105**(31), 11002–11007.

Boulin, T., Etchberger, J. F., and Hobert, O. (2006). Reporter gene fusions. *WormBook* 1–23.

Boulin, T., Gielen, M., Richmond, J. E., Williams, D. C., Paoletti, P., and Bessereau, J. L. (2008). Eight genes are required for functional reconstitution of the *Caenorhabditis elegans* levamisole-sensitive acetylcholine receptor. *Proc. Natl. Acad. Sci. USA* **105**(47), 18590–18595.

Brading, A. F., and Caldwell, P. C. (1971). The resting membrane potential of the somatic muscle cells of *Ascaris lumbricoides*. *J. Physiol.* **217**, 605–624.

Brenner, S. (1974). The genetics of *Caenorhabditis elegans*. *Genetics* **77**(1), 71–94.

Brockie, P. J., and Maricq, A. V. (2006). Ionotropic glutamate receptors: Genetics, behavior and electrophysiology. *WormBook* 1–16.

Burbea, M., Dreier, L., Dittman, J. S., Grunwald, M. E., and Kaplan, J. M. (2002). Ubiquitin and AP180 regulate the abundance of GLR-1 glutamate receptors at postsynaptic elements in *C. elegans*. *Neuron* **35**(1), 107–120.

Cabantous, S., Terwilliger, T. C., and Waldo, G. S. (2005). Protein tagging and detection with engineered self-assembling fragments of green fluorescent protein. *Nat. Biotechnol.* **23**(1), 102–107.

Cebria, F. (2007). Regenerating the central nervous system: How easy for planarians! *Dev. Genes Evol.* **217**(11–12), 733–748.

Chalasani, S. H., Chronis, N., Tsunozaki, M., Gray, J. M., Ramot, D., Goodman, M. B., and Bargmann, C. I. (2007). Dissecting a circuit for olfactory behaviour in *Caenorhabditis elegans*. *Nature* **450**(7166), 63–70.

Chalfie, M., Tu, Y., Euskirchen, G., Ward, W. W., and Prasher, D. C. (1994). Green fluorescent protein as a marker for gene expression. *Science* **263**(5148), 802–805.

Chen, Z. L., Yu, W. M., and Strickland, S. (2007). Peripheral regeneration. *Annu. Rev. Neurosci.* **30**, 209–233.

Chi, C. A., Clark, D. A., Lee, S., Biron, D., Luo, L., Gabel, C. V., Brown, J., Sengupta, P., and Samuel, A. D. (2007). Temperature and food mediate long-term thermotactic behavioral plasticity by association-independent mechanisms in *C. elegans*. *J. Exp. Biol.* **210**(Pt. 22), 4043–4052.

Ch'ng, Q., Sieburth, D., and Kaplan, J. M. (2008). Profiling synaptic proteins identifies regulators of insulin secretion and lifespan. *PLoS Genet.* **4**(11), e1000283.

Chronis, N., Zimmer, M., and Bargmann, C. I. (2007). Microfluidics for *in vivo* imaging of neuronal and behavioral activity in *Caenorhabditis elegans*. *Nat. Methods* **4**(9), 727–731.

Chung, S. H., Clark, D. A., Gabel, C. V., Mazur, E., and Samuel, A. D. (2006). The role of the AFD neuron in *C. elegans* thermotaxis analyzed using femtosecond laser ablation. *BMC Neurosci.* **7**, 30.

Chung, K., Crane, M. M., and Lu, H. (2008). Automated on-chip rapid microscopy, phenotyping and sorting of *C. elegans*. *Nat. Methods* **5**(7), 637–643.

Clark, D. A., Biron, D., Sengupta, P., and Samuel, A. D. (2006). The AFD sensory neurons encode multiple functions underlying thermotactic behavior in *Caenorhabditis elegans*. *J. Neurosci.* **26** (28), 7444–7451.

Clark, D. A., Gabel, C. V., Gabel, H., and Samuel, A. D. (2007a). Temporal activity patterns in thermosensory neurons of freely moving *Caenorhabditis elegans* encode spatial thermal gradients. *J. Neurosci.* **27**(23), 6083–6090.

Clark, D. A., Gabel, C. V., Lee, T. M., and Samuel, A. D. (2007b). Short-term adaptation and temporal processing in the cryophilic response of *Caenorhabditis elegans*. *J. Neurophysiol.* **97**(3), 1903–1910.

Cognet, L., Groc, L., Lounis, B., and Choquet, D. (2006). Multiple routes for glutamate receptor trafficking: Surface diffusion and membrane traffic cooperate to bring receptors to synapses. *Sci STKE* **2006**(327), pe13.

Crump, J. G., Zhen, M., Jin, Y., and Bargmann, C. I. (2001). The SAD-1 kinase regulates presynaptic vesicle clustering and axon termination. *Neuron* **29**(1), 115–129.

Cui, X., Lee, L. M., Heng, X., Zhong, W., Sternberg, P. W., Psaltis, D., and Yang, C. (2008). Lensless high-resolution on-chip optofluidic microscopes for Caenorhabditis elegans and cell imaging. Proc. Natl. Acad. Sci. USA 105(31), 10670–10675.

Dal Santo, P., Logan, M. A., Chisholm, A. D., and Jorgensen, E. M. (1999). The inositol trisphosphate receptor regulates a 50-second behavioral rhythm in C. elegans. Cell 98(6), 757–767.

Dittman, J. S., and Kaplan, J. M. (2006). Factors regulating the abundance and localization of synaptobrevin in the plasma membrane. Proc. Natl. Acad. Sci. USA 103(30), 11399–11404.

Dittman, J. S., and Kaplan, J. M. (2008). Behavioral impact of neurotransmitter-activated G-protein-coupled receptors: Muscarinic and GABAB receptors regulate Caenorhabditis elegans locomotion. J. Neurosci. 28(28), 7104–7112.

Dreier, L., Burbea, M., and Kaplan, J. M. (2005). LIN-23-mediated degradation of beta-catenin regulates the abundance of GLR-1 glutamate receptors in the ventral nerve cord of C. elegans. Neuron 46(1), 51–64.

Dybbs, M., Ngai, J., and Kaplan, J. M. (2005). Using microarrays to facilitate positional cloning: Identification of tomosyn as an inhibitor of neurosecretion. PLoS Genet. 1(1), 6–16.

Fares, H., and Greenwald, I. (2001a). Genetic analysis of endocytosis in Caenorhabditis elegans: Coelomocyte uptake defective mutants. Genetics 159(1), 133–145.

Fares, H., and Greenwald, I. (2001b). Regulation of endocytosis by CUP-5, the Caenorhabditis elegans mucolipin-1 homolog. Nat. Genet. 28(1), 64–68.

Faumont, S., and Lockery, S. R. (2006). The awake behaving worm: Simultaneous imaging of neuronal activity and behavior in intact animals at millimeter scale. J. Neurophysiol. 95(3), 1976–1981.

Feinberg, E. H., Vanhoven, M. K., Bendesky, A., Wang, G., Fetter, R. D., Shen, K., and Bargmann, C. I. (2008). GFP reconstitution across synaptic partners (GRASP) defines cell contacts and synapses in living nervous systems. Neuron 57(3), 353–363.

Fernandez-Alfonso, T., Kwan, R., and Ryan, T. A. (2006). Synaptic vesicles interchange their membrane proteins with a large surface reservoir during recycling. Neuron 51(2), 179–186.

Francis, M. M., Mellem, J. E., and Maricq, A. V. (2003). Bridging the gap between genes and behavior: Recent advances in the electrophysiological analysis of neural function in Caenorhabditis elegans. Trends Neurosci. 26(2), 90–99.

Francis, M. M., Evans, S. P., Jensen, M., Madsen, D. M., Mancuso, J., Norman, K. R., and Maricq, A. V. (2005). The Ror receptor tyrosine kinase CAM-1 is required for ACR-16-mediated synaptic transmission at the C. elegans neuromuscular junction. Neuron 46(4), 581–594.

Gabel, C. V., Gabel, H., Pavlichin, D., Kao, A., Clark, D. A., and Samuel, A. D. (2007). Neural circuits mediate electrosensory behavior in Caenorhabditis elegans. J. Neurosci. 27(28), 7586–7596.

Gabel, C. V., Antonie, F., Chuang, C. F., Samuel, A. D., and Chang, C. (2008). Distinct cellular and molecular mechanisms mediate initial axon development and adult-stage axon regeneration in C. elegans. Development 135(6), 1129–1136.

Gally, C., Eimer, S., Richmond, J. E., and Bessereau, J. L. (2004). A transmembrane protein required for acetylcholine receptor clustering in Caenorhabditis elegans. Nature 431(7008), 578–582.

Giles, A. C., Rose, J. K., and Rankin, C. H. (2006). Investigations of learning and memory in Caenorhabditis elegans. Int. Rev. Neurobiol. 69, 37–71.

Goodman, M. B., Hall, D. H., Avery, L., and Lockery, S. R. (1998). Active currents regulate sensitivity and dynamic range in C. elegans neurons. Neuron 20(4), 763–772.

Gottschalk, A., and Schafer, W. R. (2006). Visualization of integral and peripheral cell surface proteins in live Caenorhabditis elegans. J. Neurosci. Methods 154(1–2), 68–79.

Gottschalk, A., Almedom, R. B., Schedletzky, T., Anderson, S. D., Yates, J. R., III, and Schafer, W. R. (2005). Identification and characterization of novel nicotinic receptor-associated proteins in Caenorhabditis elegans. EMBO J. 24(14), 2566–2578.

Gracheva, E. O., Burdina, A. O., Holgado, A. M., Berthelot-Grosjean, M., Ackley, B. D., Hadwiger, G., Nonet, M. L., Weimer, R. M., and Richmond, J. E. (2006). Tomosyn inhibits synaptic vesicle priming in *Caenorhabditis elegans*. *PLoS Biol.* **4**(8), e261.

Gracheva, E. O., Burdina, A. O., Touroutine, D., Berthelot-Grosjean, M., Parekh, H., and Richmond, J. E. (2007). Tomosyn negatively regulates CAPS-dependent peptide release at *Caenorhabditis elegans* synapses. *J. Neurosci.* **27**(38), 10176–10184.

Gray, J. M., Hill, J. J., and Bargmann, C. I. (2005). A circuit for navigation in *Caenorhabditis elegans*. *Proc. Natl. Acad. Sci. USA* **102**(9), 3184–3191.

Greer, P. L., and Greenberg, M. E. (2008). From synapse to nucleus: Calcium-dependent gene transcription in the control of synapse development and function. *Neuron* **59**(6), 846–860.

Guo, S. X., Bourgeois, F., Chokshi, T., Durr, N. J., Hilliard, M. A., Chronis, N., and Ben-Yakar, A. (2008). Femtosecond laser nanoaxotomy lab-on-a-chip for *in vivo* nerve regeneration studies. *Nat. Methods* **5**(6), 531–533.

Hedgecock, E. M., and Russell, R. L. (1975). Normal and mutant thermotaxis in the nematode *Caenorhabditis elegans*. *Proc. Natl. Acad. Sci. USA* **72**(10), 4061–4065.

Helmchen, F., and Waters, J. (2002). Ca^{2+} imaging in the mammalian brain *in vivo*. *Eur. J. Pharmacol.* **447**(2–3), 119–129.

Hiatt, S. M., Shyu, Y. J., Duren, H. M., and Hu, C. D. (2008). Bimolecular fluorescence complementation (bifc) analysis of protein interactions in *Caenorhabditis elegans*. *Methods* **45**(3), 185–191.

Higashijima, S., Masino, M. A., Mandel, G., and Fetcho, J. R. (2003). Imaging neuronal activity during zebrafish behavior with a genetically encoded calcium indicator. *J. Neurophysiol.* **90**(6), 3986–3997.

Higley, M. J., and Sabatini, B. L. (2008). Calcium signaling in dendrites and spines: Practical and functional considerations. *Neuron* **59**(6), 902–913.

Hille, B. (2001). Ion Channels of Excitable Membranes p. 814. Sinauer Associates, Sunderland, MA.

Hilliard, M. A., Apicella, A. J., Kerr, R., Suzuki, H., Bazzicalupo, P., and Schafer, W. R. (2005). *In vivo* imaging of C. elegans ASH neurons: Cellular response and adaptation to chemical repellents. *EMBO J.* **24**(1), 63–72.

Hobert, O., and Loria, P. (2006). Uses of GFP in *Caenorhabditis elegans*. *Methods Biochem. Anal.* **47**, 203–226.

Holz, R. W., and Axelrod, D. (2002). Localization of phosphatidylinositol 4,5-P(2) important in exocytosis and a quantitative analysis of chromaffin granule motion adjacent to the plasma membrane. *Ann. N. Y. Acad. Sci.* **971**, 232–243.

Hulme, S. E., Shevkoplyas, S. S., Apfeld, J., Fontana, W., and Whitesides, G. M. (2007). A microfabricated array of clamps for immobilizing and imaging C. elegans. *Lab Chip* **7**(11), 1515–1523.

Hynes, T. R., Tang, L., Mervine, S. M., Sabo, J. L., Yost, E. A., Devreotes, P. N., and Berlot, C. H. (2004). Visualization of G protein betagamma dimers using bimolecular fluorescence complementation demonstrates roles for both beta and gamma in subcellular targeting. *J. Biol. Chem.* **279** (29), 30279–30286.

Jahn, R., Lang, T., and Sudhof, T. C. (2003). Membrane fusion. *Cell* **112**(4), 519–533.

Jin, Y. (2005). Synaptogenesis. *WormBook* 1–11.

Johnston, D., and Wu, S. M. S. (1995). "Foundations of Cellular Neurophysiology." The MIT Press, Cambridge, MA. p. 710.

Jones, A. K., and Sattelle, D. B. (2004). Functional genomics of the nicotinic acetylcholine receptor gene family of the nematode *Caenorhabditis elegans*. *Bioessays* **26**(1), 39–49.

Jorgensen, E. M., Hartwieg, E., Schuske, K., Nonet, M. L., Jin, Y., and Horvitz, H. R. (1995). Defective recycling of synaptic vesicles in synaptotagmin mutants of *Caenorhabditis elegans*. *Nature* **378**(6553), 196–199.

Jospin, M., Jacquemond, V., Mariol, M. C., Segalat, L., and Allard, B. (2002). The L-type voltage-dependent Ca^{2+} channel EGL-19 controls body wall muscle function in Caenorhabditis elegans. J. Cell Biol. 159(2), 337–348.

Juo, P., and Kaplan, J. M. (2004). The anaphase-promoting complex regulates the abundance of GLR-1 glutamate receptors in the ventral nerve cord of C. elegans. Curr. Biol. 14(22), 2057–2062.

Kano, T., Brockie, P. J., Sassa, T., Fujimoto, H., Kawahara, Y., Iino, Y., Mellem, J. E., Madsen, D. M., Hosono, R., and Maricq, A. V. (2008). Memory in Caenorhabditis elegans is mediated by NMDA-type ionotropic glutamate receptors. Curr. Biol. 18(13), 1010–1015.

Kerr, R. A. (2006). Imaging the activity of neurons and muscles. WormBook 1–13.

Kerr, R. A., and Schafer, W. R. (2006). Intracellular Ca^{2+} imaging in C. elegans. Methods Mol. Biol. 351, 253–264.

Kerr, R., Lev-Ram, V., Baird, G., Vincent, P., Tsien, R. Y., and Schafer, W. R. (2000). Optical imaging of calcium transients in neurons and pharyngeal muscle of C. elegans. Neuron 26(3), 583–594.

Koushika, S. P., Richmond, J. E., Hadwiger, G., Weimer, R. M., Jorgensen, E. M., and Nonet, M. L. (2001). A post-docking role for active zone protein Rim. Nat. Neurosci. 4(10), 997–1005.

Kuhara, A., Okumura, M., Kimata, T., Tanizawa, Y., Takano, R., Kimura, K. D., Inada, H., Matsumoto, K., and Mori, I. (2008). Temperature sensing by an olfactory neuron in a circuit controlling behavior of C. elegans. Science 320(5877), 803–807.

Lackner, M. R., Nurrish, S. J., and Kaplan, J. M. (1999). Facilitation of synaptic transmission by EGL-30 Gqalpha and EGL-8 plcbeta: DAG binding to UNC-13 is required to stimulate acetylcholine release. Neuron 24(2), 335–346.

Lewis, J. A., Wu, C. H., Berg, H., and Levine, J. H. (1980a). The genetics of levamisole resistance in the nematode Caenorhabditis elegans. Genetics 95(4), 905–928.

Lewis, J. A., Wu, C. H., Levine, J. H., and Berg, H. (1980b). Levamisole-resistant mutants of the nematode Caenorhabditis elegans appear to lack pharmacological acetylcholine receptors. Neuroscience 5(6), 967–989.

Li, C., and Kim, K. (2008). Neuropeptides. WormBook 1–36.

Li, Z., and Murthy, V. N. (2001). Visualizing postendocytic traffic of synaptic vesicles at hippocampal synapses. Neuron 31(4), 593–605.

Li, C., Nelson, L. S., Kim, K., Nathoo, A., and Hart, A. C. (1999). Neuropeptide gene families in the nematode Caenorhabditis elegans. Ann. N. Y. Acad. Sci. 897, 239–252.

Li, W., Feng, Z., Sternberg, P. W., and Xu, X. Z. (2006). A C. elegans stretch receptor neuron revealed by a mechanosensitive trp channel homologue. Nature 440(7084), 684–687.

Lin, D. T., and Huganir, R. L. (2007). PICK1 and phosphorylation of the glutamate receptor 2 (glur2) AMPA receptor subunit regulates glur2 recycling after NMDA receptor-induced internalization. J. Neurosci. 27(50), 13903–13908.

Liu, Q., Chen, B., Yankova, M., Morest, D. K., Maryon, E., Hand, A. R., Nonet, M. L., and Wang, Z. W. (2005). Presynaptic ryanodine receptors are required for normal quantal size at the Caenorhabditis elegans neuromuscular junction. J. Neurosci. 25(29), 6745–6754.

Lockery, S. R., Lawton, K. J., Doll, J. C., Faumont, S., Coulthard, S. M., Thiele, T. R., Chronis, N., McCormick, K. E., Goodman, M. B., and Pruitt, B. L. (2008). Artificial dirt: Microfluidic substrates for nematode neurobiology and behavior. J. Neurophysiol. 99(6), 3136–3143.

Madison, J. M., Nurrish, S., and Kaplan, J. M. (2005). UNC-13 interaction with syntaxin is required for synaptic transmission. Curr. Biol. 15(24), 2236–2242.

Magliery, T. J., Wilson, C. G., Pan, W., Mishler, D., Ghosh, I., Hamilton, A. D., and Regan, L. (2005). Detecting protein–protein interactions with a green fluorescent protein fragment reassembly trap: Scope and mechanism. J. Am. Chem. Soc. 127(1), 146–157.

Malinow, R. (2003). AMPA receptor trafficking and long-term potentiation. Philos. Trans. R. Soc. Lond. B Biol. Sci. 358(1432), 707–714.

Malinow, R., and Malenka, R. C. (2002). AMPA receptor trafficking and synaptic plasticity. *Annu. Rev. Neurosci.* **25,** 103–126.

Marella, S., Fischler, W., Kong, P., Asgarian, S., Rueckert, E., and Scott, K. (2006). Imaging taste responses in the fly brain reveals a functional map of taste category and behavior. *Neuron* **49**(2), 285–295.

Margeta, M. A., Shen, K., and Grill, B. (2008). Building a synapse: Lessons on synaptic specificity and presynaptic assembly from the nematode *C. elegans*. *Curr. Opin. Neurobiol.* **18**(1), 69–76.

Marza, E., Long, T., Saiardi, A., Sumakovic, M., Eimer, S., Hall, D. H., and Lesa, G. M. (2008). Polyunsaturated fatty acids influence synaptojanin localization to regulate synaptic vesicle recycling. *Mol. Biol. Cell* **19**(3), 833–842.

McEwen, J. M., Madison, J. M., Dybbs, M., and Kaplan, J. M. (2006). Antagonistic regulation of synaptic vesicle priming by Tomosyn and UNC-13. *Neuron* **51**(3), 303–315.

McGee, A. W., and Strittmatter, S. M. (2003). The Nogo-66 receptor: Focusing myelin inhibition of axon regeneration. *Trends Neurosci.* **26**(4), 193–198.

Mellem, J. E., Brockie, P. J., Madsen, D. M., and Maricq, A. V. (2008). Action potentials contribute to neuronal signaling in *C. elegans*. *Nat. Neurosci.* **11**(8), 865–867.

Micheva, K. D., Holz, R. W., and Smith, S. J. (2001). Regulation of presynaptic phosphatidylinositol 4,5-biphosphate by neuronal activity. *J. Cell Biol.* **154**(2), 355–368.

Miesenbock, G., De Angelis, D. A., and Rothman, J. E. (1998). Visualizing secretion and synaptic transmission with pH-sensitive green fluorescent proteins. *Nature* **394**(6689), 192–195.

Miyawaki, A. (2003). Fluorescence imaging of physiological activity in complex systems using GFP-based probes. *Curr. Opin. Neurobiol.* **13**(5), 591–596.

Miyawaki, A. (2005). Innovations in the imaging of brain functions using fluorescent proteins. *Neuron* **48**(2), 189–199.

Mori, I., Sasakura, H., and Kuhara, A. (2007). Worm thermotaxis: A model system for analyzing thermosensation and neural plasticity. *Curr. Opin. Neurobiol.* **17**(6), 712–719.

Motley, A. M., Berg, N., Taylor, M. J., Sahlender, D. A., Hirst, J., Owen, D. J., and Robinson, M. S. (2006). Functional analysis of AP-2 alpha and mu2 subunits. *Mol. Biol. Cell* **17**(12), 5298–5308.

Murfitt, R. R., Vogel, K., and Sanadi, D. R. (1976). Characterization of the mitochondria of the free-living nematode, *Caenorhabditis elegans*. *Comp. Biochem. Physiol. B* **53**(4), 423–430.

Nakai, J., Ohkura, M., and Imoto, K. (2001). A high signal-to-noise Ca(2+) probe composed of a single green fluorescent protein. *Nat. Biotechnol.* **19**(2), 137–141.

Nonet, M. L. (1999). Visualization of synaptic specializations in live *C. elegans* with synaptic vesicle protein-GFP fusions. *J. Neurosci. Methods* **89**(1), 33–40.

Nonet, M. L., Holgado, A. M., Brewer, F., Serpe, C. J., Norbeck, B. A., Holleran, J., Wei, L., Hartwieg, E., Jorgensen, E. M., and Alfonso, A. (1999). UNC-11, a *Caenorhabditis elegans* AP180 homologue, regulates the size and protein composition of synaptic vesicles. *Mol. Biol. Cell* **10**(7), 2343–2360.

Nurrish, S., Segalat, L., and Kaplan, J. M. (1999). Serotonin inhibition of synaptic transmission: Galpha(0) decreases the abundance of UNC-13 at release sites. *Neuron* **24**(1), 231–242.

Park, S., Hwang, H., Nam, S. W., Martinez, F., Austin, R. H., and Ryu, W. S. (2008). Enhanced *Caenorhabditis elegans* locomotion in a structured microfluidic environment. *PLoS ONE* **3**(6), e2550.

Patel, M. R., Lehrman, E. K., Poon, V. Y., Crump, J. G., Zhen, M., Bargmann, C. I., and Shen, K. (2006). Hierarchical assembly of presynaptic components in defined *C. elegans* synapses. *Nat. Neurosci.* **9**(12), 1488–1498.

Pierce-Shimomura, J. T., Faumont, S., Gaston, M. R., Pearson, B. J., and Lockery, S. R. (2001). The homeobox gene lim-6 is required for distinct chemosensory representations in *C. elegans*. *Nature* **410**(6829), 694–698.

Pologruto, T. A., Yasuda, R., and Svoboda, K. (2004). Monitoring neural activity and [Ca^{2+}] with genetically encoded Ca^{2+} indicators. *J. Neurosci.* **24**(43), 9572–9579.

Ramot, D., MacInnis, B. L., and Goodman, M. B. (2008a). Bidirectional temperature-sensing by a single thermosensory neuron in C. *elegans. Nat. Neurosci.* **11**(8), 908–915.

Ramot, D., MacInnis, B. L., Lee, H. C., and Goodman, M. B. (2008b). Thermotaxis is a robust mechanism for thermoregulation in *Caenorhabditis elegans* nematodes. *J. Neurosci.* **28**(47), 12546–12557.

Rand, J. B. (2007). Acetylcholine. *WormBook* 1–21.

Rao, G. N., Kulkarni, S. S., Koushika, S. P., and Rau, K. R. (2008). *In vivo* nanosecond laser axotomy: Cavitation dynamics and vesicle transport. *Opt. Express* **16**(13), 9884–9894.

Rappoport, J. Z., and Simon, S. M. (2008). A functional GFP fusion for imaging clathrin-mediated endocytosis. *Traffic* **9**(8), 1250–1255.

Regehr, W. G., and Tank, D. W. (1991). Selective fura-2 loading of presynaptic terminals and nerve cell processes by local perfusion in mammalian brain slice. *J. Neurosci. Methods* **37**(2), 111–119.

Regehr, W. G., and Tank, D. W. (1994). Dendritic calcium dynamics. *Curr. Opin. Neurobiol.* **4**(3), 373–382.

Reiff, D. F., Ihring, A., Guerrero, G., Isacoff, E. Y., Joesch, M., Nakai, J., and Borst, A. (2005). *In vivo* performance of genetically encoded indicators of neural activity in flies. *J. Neurosci.* **25**(19), 4766–4778.

Richmond, J. E., and Jorgensen, E. M. (1999). One GABA and two acetylcholine receptors function at the C. *elegans* neuromuscular junction. *Nat. Neurosci.* **2**(9), 791–797.

Richmond, J. E., Davis, W. S., and Jorgensen, E. M. (1999). UNC-13 is required for synaptic vesicle fusion in C. *elegans. Nat. Neurosci.* **2**(11), 959–964.

Rohde, C. B., Zeng, F., Gonzalez-Rubio, R., Angel, M., and Yanik, M. F. (2007). Microfluidic system for on-chip high-throughput whole-animal sorting and screening at subcellular resolution. *Proc. Natl. Acad. Sci. USA* **104**(35), 13891–13895.

Rongo, C., Whitfield, C. W., Rodal, A., Kim, S. K., and Kaplan, J. M. (1998). LIN-10 is a shared component of the polarized protein localization pathways in neurons and epithelia. *Cell* **94**(6), 751–759.

Rowland, A. M., Richmond, J. E., Olsen, J. G., Hall, D. H., and Bamber, B. A. (2006). Presynaptic terminals independently regulate synaptic clustering and autophagy of GABAA receptors in *Caenorhabditis elegans. J. Neurosci.* **26**(6), 1711–1720.

Ryu, W. S., and Samuel, A. D. (2002). Thermotaxis in *Caenorhabditis elegans* analyzed by measuring responses to defined thermal stimuli. *J. Neurosci.* **22**(13), 5727–5733.

Salkoff, L., Wei, A. D., Baban, B., Butler, A., Fawcett, G., Ferreira, G., and Santi, C. M. (2005). Potassium channels in C. *elegans. WormBook* 1–15.

Samuel, A. D., Silva, R. A., and Murthy, V. N. (2003). Synaptic activity of the AFD neuron in *Caenorhabditis elegans* correlates with thermotactic memory. *J. Neurosci.* **23**(2), 373–376.

Sankaranarayanan, S., and Ryan, T. A. (2000). Real-time measurements of vesicle-SNARE recycling in synapses of the central nervous system. *Nat. Cell Biol.* **2**(4), 197–204.

Sankaranarayanan, S., De Angelis, D., Rothman, J. E., and Ryan, T. A. (2000). The use of phluorins for optical measurements of presynaptic activity. *Biophys. J.* **79**(4), 2199–2208.

Schaefer, H., and Rongo, C. (2006). KEL-8 is a substrate receptor for CUL3-dependent ubiquitin ligase that regulates synaptic glutamate receptor turnover. *Mol. Biol. Cell* **17**(3), 1250–1260.

Schafer, W. R. (2005). Egg-laying. *WormBook* 1–7.

Schiller, P. H. (1992). The ON and OFF channels of the visual system. *Trends Neurosci.* **15**(3), 86–92.

Schuske, K. R., Richmond, J. E., Matthies, D. S., Davis, W. S., Runz, S., Rube, D. A., van der Bliek, A. M., and Jorgensen, E. M. (2003). Endophilin is required for synaptic vesicle endocytosis by localizing synaptojanin. *Neuron* **40**(4), 749–762.

Shapira, M., Zhai, R. G., Dresbach, T., Bresler, T., Torres, V. I., Gundelfinger, E. D., Ziv, N. E., and Garner, C. C. (2003). Unitary assembly of presynaptic active zones from Piccolo-Bassoon transport vesicles. *Neuron* **38**(2), 237–252.

Sieburth, D., Ch'ng, Q., Dybbs, M., Tavazoie, M., Kennedy, S., Wang, D., Dupuy, D., Rual, J. F., Hill, D. E., Vidal, M., Ruvkun, G., and Kaplan, J. M. (2005). Systematic analysis of genes required for synapse structure and function. *Nature* **436**(7050), 510–517.

Sieburth, D., Madison, J. M., and Kaplan, J. M. (2007). PKC-1 regulates secretion of neuropeptides. *Nat. Neurosci.* **10**(1), 49–57.

Simon, D. J., Madison, J. M., Conery, A. L., Thompson-Peer, K. L., Soskis, M., Ruvkun, G. B., Kaplan, J. M., and Kim, J. K. (2008). The microrna mir-1 regulates a MEF-2-dependent retrograde signal at neuromuscular junctions. *Cell* **133**(5), 903–915.

Speese, S., Petrie, M., Schuske, K., Ailion, M., Ann, K., Iwasaki, K., Jorgensen, E. M., and Martin, T. F. (2007). UNC-31 (CAPS) is required for dense-core vesicle but not synaptic vesicle exocytosis in *Caenorhabditis elegans*. *J. Neurosci.* **27**(23), 6150–6162.

Stauffer, T. P., Ahn, S., and Meyer, T. (1998). Receptor-induced transient reduction in plasma membrane ptdins(4,5)P2 concentration monitored in living cells. *Curr. Biol.* **8**(6), 343–346.

Stretton, A., Donmoyer, J., Davis, R., Meade, J., Cowden, C., and Sithigorngul, P. (1992). Motor behavior and motor nervous system function in the nematode *Ascaris suum*. *J. Parasitol.* **78**, 206–214.

Sulston, J. E., and Horvitz, H. R. (1977). Post-embryonic cell lineages of the nematode, *Caenorhabditis elegans*. *Dev. Biol.* **56**(1), 110–156.

Suzuki, H., Kerr, R., Bianchi, L., Frokjaer-Jensen, C., Slone, D., Xue, J., Gerstbrein, B., Driscoll, M., and Schafer, W. R. (2003). In vivo imaging of *C. elegans* mechanosensory neurons demonstrates a specific role for the MEC-4 channel in the process of gentle touch sensation. *Neuron* **39**(6), 1005–1017.

Suzuki, H., Thiele, T. R., Faumont, S., Ezcurra, M., Lockery, S. R., and Schafer, W. R. (2008). Functional asymmetry in *Caenorhabditis elegans* taste neurons and its computational role in chemotaxis. *Nature* **454**(7200), 114–117.

Takamori, S., Holt, M., Stenius, K., Lemke, E. A., Gronborg, M., Riedel, D., Urlaub, H., Schenck, S., Brugger, B., Ringler, P., Muller, S. A., Rammner, B., *et al.* (2006). Molecular anatomy of a trafficking organelle. *Cell* **127**(4), 831–846.

Tao-Cheng, J. H. (2006). Activity-related redistribution of presynaptic proteins at the active zone. *Neuroscience* **141**(3), 1217–1224.

Touroutine, D., Fox, R. M., Von Stetina, S. E., Burdina, A., Miller, D. M., III, and Richmond, J. E. (2005). acr-16 encodes an essential subunit of the levamisole-resistant nicotinic receptor at the *Caenorhabditis elegans* neuromuscular junction. *J. Biol. Chem.* **280**(29), 27013–27021.

Tsien, R. Y. (1998). The green fluorescent protein. *Annu. Rev. Biochem.* **67**, 509–544.

Wang, Z. W., Saifee, O., Nonet, M. L., and Salkoff, L. (2001). SLO-1 potassium channels control quantal content of neurotransmitter release at the *C. elegans* neuromuscular junction. *Neuron* **32**(5), 867–881.

Weimer, R. M., and Jorgensen, E. M. (2003). Controversies in synaptic vesicle exocytosis. *J. Cell Sci.* **116**(Pt. 18), 3661–3666.

Weimer, R. M., Richmond, J. E., Davis, W. S., Hadwiger, G., Nonet, M. L., and Jorgensen, E. M. (2003). Defects in synaptic vesicle docking in unc-18 mutants. *Nat. Neurosci.* **6**(10), 1023–1030.

White, J. G., Southgate, E., Thomson, J. N., and Brenner, S. (1976). The structure of the ventral nerve cord of *Caenorhabditis elegans*. *Philos. Trans. R. Soc. Lond. B Biol. Sci.* **275**(938), 327–348.

White, J. G., Horvitz, H. R., and Sulston, J. E. (1982). Neurone differentiation in cell lineage mutants of *Caenorhabditis elegans*. *Nature* **297**(5867), 584–587.

White, J. G., Southgate, E., Thomson, J. N., and Brenner, S. (1986). The structure of the nervous system of *Caenorhabditis elegans*. *Philos. Trans. R. Soc. Lond. B Biol. Sci.* **314**, 1–340.

Whittaker, A. J., and Sternberg, P. W. (2004). Sensory processing by neural circuits in *Caenorhabditis elegans*. *Curr. Opin. Neurobiol.* **14**(4), 450–456.

Wienisch, M., and Klingauf, J. (2006). Vesicular proteins exocytosed and subsequently retrieved by compensatory endocytosis are nonidentical. *Nat. Neurosci.* **9**(8), 1019–1027.

Wu, Z., Ghosh-Roy, A., Yanik, M. F., Zhang, J. Z., Jin, Y., and Chisholm, A. D. (2007). *Caenorhabditis elegans* neuronal regeneration is influenced by life stage, ephrin signaling, and synaptic branching. *Proc. Natl. Acad. Sci. USA* **104**(38), 15132–15137.

Yanik, M. F., Cinar, H., Cinar, H. N., Chisholm, A. D., Jin, Y., and Ben-Yakar, A. (2004). Neurosurgery: Functional regeneration after laser axotomy. *Nature* **432**(7019), 822.

Zaccolo, M., De Giorgi, F., Cho, C. Y., Feng, L., Knapp, T., Negulescu, P. A., Taylor, S. S., Tsien, R. Y., and Pozzan, T. (2000). A genetically encoded, fluorescent indicator for cyclic AMP in living cells. *Nat. Cell Biol.* **2**(1), 25–29.

Zhang, Y., Grant, B., and Hirsh, D. (2001). RME-8, a conserved J-domain protein, is required for endocytosis in *Caenorhabditis elegans*. *Mol. Biol. Cell* **12**(7), 2011–2021.

Zhang, S., Ma, C., and Chalfie, M. (2004). Combinatorial marking of cells and organelles with reconstituted fluorescent proteins. *Cell* **119**(1), 137–144.

Zhang, M., Chung, S. H., Fang-Yen, C., Craig, C., Kerr, R. A., Suzuki, H., Samuel, A. D., Mazur, E., and Schafer, W. R. (2008). A self-regulating feed-forward circuit controlling C. *elegans* egg-laying behavior. *Curr. Biol.* **18**(19), 1445–1455.

Zhen, M., Huang, X., Bamber, B., and Jin, Y. (2000). Regulation of presynaptic terminal organization by C. *elegans* RPM-1, a putative guanine nucleotide exchanger with a RING-H2 finger domain. *Neuron* **26**(2), 331–343.

3

Mapping and Manipulating Neural Circuits in the Fly Brain

Julie H. Simpson
HHMI Janelia Farm Research Campus, Asburn, VA, USA

I. Introduction
  A. Genes for behavior
  B. Neurons for behavior
  C. Anatomy and stereotypy
II. Spatial Targeting of Neuron Types
III. Imaging Neurons
IV. Functional Imaging: Watching Neuronal Activity
  A. Voltage sensors
  B. Genetically encoded calcium indicators
V. Control of Neural Activity
  A. Cell killers
  B. Synaptic vesicle blockers
  C. Electrical blockers
  D. Neuronal activators
  E. Light-based methods
  F. Caveats
VI. Quantitative Behavioral Assays
VII. Conclusions
  A. Example circuits
  B. New tools
  C. Full circle
  Acknowledgments
  References

Advances in Genetics, Vol. 65
Copyright 2009, Elsevier Inc. All rights reserved.

ABSTRACT

Drosophila is a marvelous system to study the underlying principles that govern how neural circuits govern behaviors. The scale of the fly brain (~100,000 neurons) and the complexity of the behaviors the fly can perform make it a tractable experimental model organism. In addition, 100 years and hundreds of labs have contributed to an extensive array of tools and techniques that can be used to dissect the function and organization of the fly nervous system. This review discusses both the conceptual challenges and the specific tools for a neurogenetic approach to circuit mapping in Drosophila. © 2009, Elsevier Inc.

I. INTRODUCTION

Why would you want to map neural circuits? In our quest to understand how the brain controls appropriate responses to environment and experience, we must track which neurons are connected and what jobs they do together. The wiring diagram and associated behavioral functions of neurons are prerequisites for the kind of experiments that will truly parse what the nervous system as an interconnected network does. Research for mapping neural circuits required for specific behaviors has shifted from hunting for the responsible *genes* to the responsible *neurons*. The "lesion approach," where damaged brain regions are correlated with behavioral changes, has been highly effective in vertebrates—humans, too (Damasio *et al.*, 1994)—but the spatial and temporal precision with which we can generate "lesions" in the genetic model organisms is unrivaled. This kind of targeted genetic lesion is a way to make circuit breaking into "a science of control and causality rather than a science of observation and correlation" (Holmes *et al.*, 2007). This is an exciting time to be studying neuroscience, both because of the tools available and because the trend toward multidisciplinary science and freer journal access has pushed previously under connected fields together: information from other scientific disciplines (systems neuroscience, neuroethology) and other organisms (stick insects, bees, locusts) are now informing the experiments we do in Drosophila, which has long been a genetic powerhouse for studying development and biochemical signaling pathways.

There are different kinds of information that can be gathered about neural circuits. One could collect anatomical information by labeling individual neurons or fiber tracts and determining neural shape and region-level connectivity at the light level or by electron microscopy. One could record activity in individual neurons or populations with optical reporters or electrodes. One could do careful behavioral assays and deduce what sorts of circuits must underlie particular computations from latency to response to a sensory stimulus, differences in execution/performance, or types of errors. One could screen for mutations that disrupt neural circuit formation or function. One could do injury or lesion studies to see where structural perturbations disturb behavioral output. And now we are using new technology to make genetically targeted lesions to disrupt function in specific neurons to map neural circuits directly. Figure 3.1 shows a schematic of this approach. In this section, I will discuss briefly what has been learned from these various approaches but I will devote most of the review to discussion of the tools available for generating genetically targeted disruptions in neural activity.

This review attempts to cover four areas—spatial control for targeting small groups of neurons reproducibly, visualization of the activity and connectivity of neurons, temporal control of neural activity, and behavioral assessment of defective flies. I try to give both the original references where tools were developed and examples of circuit dissection where the tools have been used particularly well. I have drawn almost exclusively from the literature on adult flies rather than larva. As a practitioner of this ilk of circuit tracing, I have used many of the reagents discussed and I have tried to inject cautionary notes based on my own experience and those of my colleagues that may not have made it into print since negative results often go undocumented. I have tried to compile best practices, appropriate controls, and areas ripe for improvement and discovery. Construct names are in **bold** for easy spotting and the ***bold italics*** text highlights references for particularly good examples of the use of the tool for circuit bashing.

Some aspects of this chapter have been ably covered in recent reviews and I refer you to them for additional information and different perspectives. Specifically, I suggest reviews of spatial control of gene expression and neuronal targeting (***Luan and White, 2007***); manipulation of neural activity (***Holmes et al., 2007***); fly circuit analysis with emphasis on electrophysiology, functional imaging, and neural computations (***Olsen and Wilson, 2008a***); vertebrate and invertebrate techniques (***Luo et al., 2008***); and genes and behavior (***Baker et al., 2001; Dickson, 2008; Vosshall, 2007***).

A. Genes for behavior

There are genetic mutations that affect behavior. Genes encode the proteins required to specify neural cell type, guide axons to their appropriate targets, drive the membrane potential changes that allow action potentials, and synthesize and release neurotransmitters. Mutagenesis screens have uncovered many of these genes. In some cases, gene expression is restricted to small groups of neurons, which gives a starting point for circuit identification. In other cases, reexpressing the missing gene in restricted subsets of neurons to show that function in these neurons is sufficient to restore normal behavior has identified the circuits underlying a given behavior.

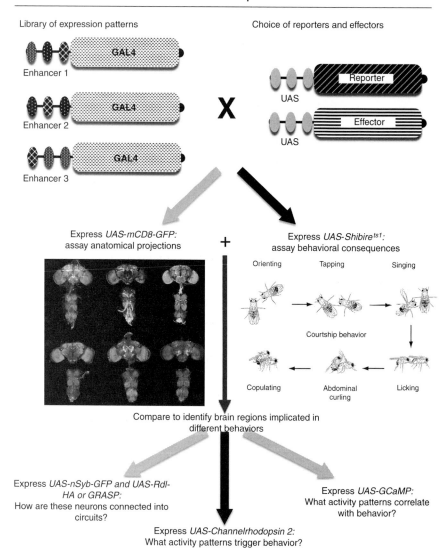

Figure 3.1. *Screening approach to identify circuit components by targeted genetic lesions.* A library of GAL4 lines is crossed to anatomical markers to determine the identity and potential connectivity of neurons. The same lines are crossed to neural activity blockers or activators and behavioral effects assayed. Lines that show similar behavior defects can be compared to look for shared neurons. The complex expression patterns can be further dissected by intersectional strategies described in Fig. 3.2. Functional imaging can also be tried to identify relevant neurons. Examples of use of this approach include Armstrong *et al.* (2006), Baker *et al.* (2007b), Gordon and Scott (2009b), Hughes and Thomas (2007), Katsov and Clandinin (2008), Kitamoto (2002), and Pitman *et al.* (2006).

There is a long history of performing radiation, chemical, or transposon mutagenesis and screening for behavioral defects. Seymour Benzer (1921–2007) was a pioneer of this approach in *Drosophila*. He identified flies defective in fast phototaxis and circadian rhythms, for example (Benzer 1967, 1973). It is tricky to identify mutants that only affect an adult behavioral phenotype since most mutants are pleiotropic, contributing to animal function during development and/or in multiple tissues. The ability to screen vast numbers of flies allowed people to obtain hypomorphic and neomorphic alleles which had more subtle effects on phenotypes (Greenspan, 1997). The Per^{long} and Per^{short} alleles of the circadian rhythm gene *Period* are examples of this (Konopka and Benzer, 1971). David Suzuki searched specifically for conditional alleles, shifting to nonpermissive temperatures in the adult to obtain specific behavioral defects (Homyk *et al.*, 1980; Suzuki *et al.*, 1971). Many of these mutations were eventually mapped to ion channel genes. Sokolowski and colleagues took advantage of a natural behavioral variant to identify the *foraging* gene in which two different alleles, neither of which is a null, affect larval feeding behavior (de Belle *et al.*, 1989). Natural variants have also been identified in population selection screens for increased lifespan and response to gravity (Lin *et al.*, 1998; Song *et al.*, 2002). Screens for the failure of the jump-escape circuit led to the cloning of an invertebrate gap junction component, the ShakingB Neural innexin (Thomas and Wyman, 1984). Screens for grooming behavior and response to ethanol have yielded mutants in adhesion molecules and cell signaling cascades (Moore *et al.*, 1998; Phillis *et al.*, 1993). How these genes contribute to the performance of these behaviors remains mysterious. Localizing which neurons require these proteins has been key for identifying the neural circuits involved.

Sometimes the genes are expressed in restricted patterns that suggest which neurons are critical for the behavior affected by mutant alleles (Hamada *et al.*, 2008; Renn *et al.*, 1999). People have used the behavioral mutants to identify the neurons participating in particular behaviors by restoring functional protein selectively—rescuing in specific cell types or time points. *CamKII* mutants are defective in the memory of bad experiences during courtship, but restoring CamKII in the mushroom bodies rescues normal memory performance (Joiner and Griffith, 1999). Flies mutant for *rutabaga* have visual memory defects that are restored by replacing rutabaga function in different layers of the fan-shaped body—as well as in some other areas of the brain (Liu *et al.*, 2006; Zars *et al.*, 2000). Expressing *taybridge* in the central complex rescues that mutant's locomotor and anatomical defects (Poeck *et al.*, 2008).

B. Neurons for behavior

Attempts to map the parts of the brain that drive behaviors go back to the days of gynandromorphs or sexual mosaics. The parts of the brain that must be genetically male to drive appropriate male courtship behaviors have been known at a rough level for decades (Hall, 1979; Hotta and Benzer, 1970; Tompkins and Hall, 1983). Laborious histological screens were done to isolate mutants with visible anatomical defects in particular brain regions; behavior analysis lead to the hypothesis that the central complex is critical for coordinated locomotion (Ilius *et al.*, 1994; Strauss and Heisenberg, 1993). Drug ablation of the mushroom bodies implicated them in memory formation and retrieval (de Belle and Heisenberg, 1994). The modern methods for targeting neural activity modifiers to specific groups of neurons and assaying behavioral consequences discussed below are a logical continuation of this tradition for circuit mapping.

In the genetic tradition, a gene is considered *necessary* for a process if null mutants disrupt the process, and *sufficient* if restoration of the gene ameliorates the phenotype. This is usually taken as proof that a given gene is the cause of a phenotype. The circuit mapping analogy is that if blocking neural activity in a group of neurons disrupts a behavior, those neurons are in some way *necessary* for the performance of that behavior. If restoring neural activity—or function of a necessary gene—specifically in a group of neurons rescues the behavior, these neurons are thought to be *sufficient*. If triggering activity in a group of neurons evokes the behavior, those neurons are capable of causing the behavior, whether they normally play this role or not. These standards of proof for implicating neurons in behavioral control are useful, but the circuits that normally drive behavior can be complex and redundant, so care should be taken to interpret the results of necessity and sufficiency experiments. With neurons as well as with genes, the expression levels and extent of rescue are rarely perfectly measured or controlled. Blocking and activating experiments in the style depicted in Fig. 3.1 are useful for identifying the component parts of neural circuits, but the way these neurons work together to drive behavior is a network property; the list of parts is necessary but not sufficient to explain circuit function.

C. Anatomy and stereotypy

Sometimes the anatomy alone gives clues about neural function and connectivity into circuits. For example, the "parts list" for the retina suggests where color comparisons could be made (Fischbach and Dittrich, 1989; Morante and Desplan, 2008). The morphology of the lobular plate tangential cells suggests that they may detect horizontal or visual motion (Joesch *et al.*, 2008; Scott *et al.*, 2002). Although there is no published quantification, there are thought to be on the order of 100,000 neurons in the adult fly nervous system: 30,000 are part of the central brain (includes the subesophogeal ganglia), 15,000 in each optic lobe, and another 15,000 in the ventral nerve cord or thoracic and abdominal ganglia. Approximately 3600 ascending and descending neurons pass through the cervical connective to connect the brain and thoracic ganglia. Neuronal cell bodies are between 2 and 5 μm in diameter, dendritic fields can span 50 μm, and neurites can extend 100 μm. In *Drosophila*, the cell bodies are located on the outside surface of the brain—the cortical rind—while the neurites project inside to form the synaptic neuropil. This region is divided into compartments by glial sheaths and axon tracts. The fly uses the canonical neurotransmitters (including acetylcholine, glutamate, GABA, histamine, dopamine, and serotonin (Bicker, 1999; Littleton and Ganetzky, 2000)) as well as tyramine, octopamine, and neuropeptides (Nassel and Homberg, 2006; Roeder, 2005; Taghert and Veenstra, 2003). How many types of neurons the fly has is the subject of much debate, but this largely depends on how one defines type: origin or lineage, transmitter type, gene expression profile, morphology, connectivity, or function. The nomenclature and descriptive anatomy of the adult fly brain is still being studied and described—no atlas or comprehensive textbook exists—although there is a serious effort underway to standardize naming conventions and disseminate this information to the research community. There remains a lot of terra incognita: brain regions whose function and connectivity is unknown.

In order for circuit mapping to be meaningful, we must ask if the circuits that drive a behavior in one individual will be similar to those that do so in another. We believe neural identity and connectivity in the fly are relatively stereotyped. The sensory projections and the circuits governing innate behaviors seem to be grossly similar from individual to individual where they have been carefully studied. For review of the olfactory projection neurons as an example, see Cachero and Jefferis (2008). The motor neurons and photoreceptors connect precisely to their targets even in the absence of neural activity (Baines *et al.*, 2001; Broadie and Bate, 1993; Hiesinger *et al.*, 2006). There are examples of morphological plasticity: the olfactory glomeruli responding to carbon dioxide expand if the flies are raised in a high CO_2 environment (Sachse *et al.*, 2007). The mushroom bodies are larger in flies raised in mixed gender groups than in those raised in isolation, and the brain areas associated with walking are larger in lab strains while those associated with flight are larger in more wild ones (Heisenberg *et al.*, 1995; Rein *et al.*, 2002). In the optic neuropils, cell size and shape can change with circadian rhythms (Pyza and Meinertzhagen, 1999). Most of these changes are due to increases in arborization or branching, and potentially increases in synaptic connections, rather than the development of entirely new circuits. Activity within a circuit might or might not be stereotypical. For instance, statistical arguments can be made from recording from many mushroom-body Kenyon cell neurons to show that their odor response profiles vary between individuals (Murthy *et al.*, 2008). Whether this affects the animals' behavioral performance is not known. Extensive work—both theoretical and experimental—in the stomatogastric system has shown that functional central pattern generators can be constructed with neurons with a range of firing properties and configurations (Prinz *et al.*, 2004; Schulz *et al.*, 2006). In the behavior assays performed in flies to date, genetically homogeneous populations tend to perform similarly. It seems reasonable to suppose that the neural circuits that underlie behavior are sufficiently stereotyped in *Drosophila* that we can learn something useful about their organizing principles.

To summarize, genes that have behavioral consequences have been identified. Unusual alleles of these genes have been more informative than nulls. These genes tend to control the development of neurons or be components of the machinery that makes them function (ion channels, SNARE proteins, enzymes, etc.). Systems for targeting neurons, rather than genes, may be more informative for sorting out principles of neural circuit organization. One can disrupt neural function to show necessity or activate neural function to determine sufficiency. The fly brain seems to be sufficiently hardwired and stereotyped that the circuits that drive a behavior should be similar in different individuals of the same genotype, allowing the deduction of general principles of how circuits organize to drive behavior.

II. SPATIAL TARGETING OF NEURON TYPES

One would like to have reproducible genetic access defined populations of neurons for circuit analysis. One can introduce exogenous genes into *Drosophila* using transposable elements (Rubin and Spradling, 1982) and generate markers for given cells by fusing an enhancer directly to an enzymatic or fluorescent reporter protein (for example, see Couto *et al.*, 2005; Wang *et al.*, 2004b). The binary **UAS-GAL4** system (Brand and Perrimon, 1993; Fischer *et al.*, 1988) uses the GAL4 transcription factor from yeast to drive transgenes of choice under the control of the UAS upstream activating sequence. This two-part system is a powerful technique for expressing different genes in the same cell types. It allows reproducible access to a given cell type to perform different manipulations. For example, one can use *ShakingB-GAL4* to express *UAS-mCD8-GFP*, a membrane-targeted green fluorescent protein, to visualize the trajectory of giant fiber neurons in one set of flies and then use the same GAL4 driver to express *UAS-Shibire^{ts1}*, a temperature-sensitive protein that blocks synaptic vesicle recycling, to disrupt neural activity in the same neurons to assay behavioral consequences in another set of flies (see Fig. 3.1). Given the stereotypy of expression from the GAL4, one can be reasonably confident that both manipulations are being done on the same population of neurons (see below—end of Section III—for a discussion of the limits of this assumption). The GAL4 line, which dictates which neurons are targeted, is referred to as the driver, while the UAS construct is called a reporter or effector. In addition to targeting different operations to the same cells, the UAS-GAL4 system also amplifies the expression level of the reporter transgene. The GAL4 system and its many uses have been reviewed often (Duffy, 2002; Phelps and Brand, 1998); the various intersectional modifications discussed below are summarized in Fig. 3.2.

It is possible to make GAL4 lines by randomly mobilizing a P-element transposon around the genome. This approach is called **enhancer trapping** and has been done extensively (Han *et al.*, 1996). The enhancer trap GAL4 lines might be expressed in the same cells as the gene whose enhancer they trap, but they might have novel patterns since they can land in the middle of enhancers or capture fragments of DNA that are serendipitously capable of driving expression. Whether the neurons labeled by a given GAL4 line constitute a "cell type" is debatable, but they are a group of cells that have at least some element of gene expression in common. P-elements have insertion site preferences (AT rich regions in the 5' ends of genes) and at this point the genome has been extensively covered with P-element inserts of GAL4. The labs of Kaiser, Ito, and Heberlein have generated large collections (Hayashi *et al.*, 2002; Manseau *et al.*, 1997; Rodan *et al.*, 2002). There are variations on the enhancer trap: a dual-headed trap can pick up enhancers from genes transcribed on either strand to increase the rate of insertions with expression patterns (Lukacsovich *et al.*, 2001). No one has yet published a large-scale GAL4 enhancer trap hop in one of the alternative transposons (piggyBac, Minos, Mariner) which have different insertion biases; this might generate new GAL4 expression patterns. The protein trap approach could also be adapted to select inserts that actually disrupt genes (Lukacsovich *et al.*, 2008; Morin *et al.*, 2001; Quinones-Coello *et al.*, 2007), which will occur less frequently with the alternative transposons.

It is also possible to design GAL4s to reflect expression of specific genes, either by knocking GAL4 into the genomic locus (Rong and Golic, 2000), as was done to make *Fruitless*^{M}-GAL4s (Demir and Dickson, 2005; Manoli *et al.*, 2005) or by using large fractions of the DNA surrounding a gene, as for TH-GAL4 (Friggi-Grelin *et al.*, 2003). The latter approach should become easier with the adoption of the bacterial artificial chromosome (BAC) insertion approach (Venken and Bellen, 2005; Venken *et al.*, 2006). It is also possible to take small pieces of DNA upstream of the coding region of interesting genes and fuse this putative regulatory DNA to GAL4 in a transformable vector (Sharma *et al.*, 2002). This designed enhancer approach has been used in the past (Hiromi *et al.*, 1985; Moses and Rubin, 1991) and a large collection of GAL4s using the regulatory regions of neural genes is being generated now (Pfeiffer *et al.*, 2008). This collection is expected to be very powerful because of the high expression level of its GAL4 vector and because all of the constructs are inserted into the same genomic locus using the PhiC1 integration system (Bischof *et al.*, 2007; Fish *et al.*, 2007; Groth *et al.*, 2004), removing position effect variation. The existing and planned GAL4 reagents come close to allowing genetic access to small intersecting subsets of neurons throughout the fly brain.

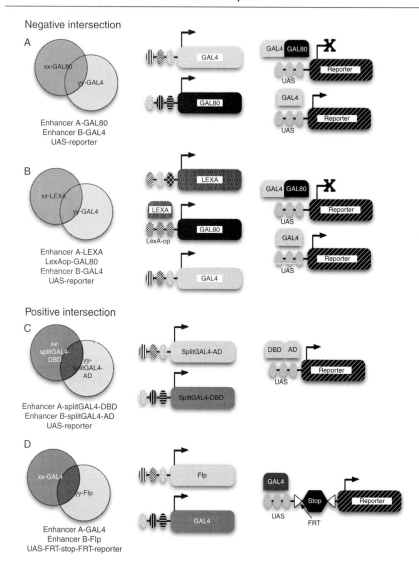

Figure 3.2. *Intersectional strategies to refine spatial expression patterns.* (A) When GAL4 and GAL80 patterns overlap, reporter expression is possible where GAL4 is made and GAL80 is not. (B) GAL80 can be expressed in response to LEXA, which may amplify its expression level. GAL4 function is restricted to the nonoverlapping region. (C) Expressing both halves of split-GAL4 in overlapping patterns restricts functional GAL4 production to the overlap. (D) Expressing a recombinase allows the removal of a stop cassette flanked with target sites. If the reporter is under UAS control as well, it is only made where the recombinase and the GAL4 expression coincide. Other strategies and combinations can be constructed from these basic building blocks.

We have not yet achieved the kind of control where one can design a regulatory sequence of transcription factor and repressor binding sites to dictate the location and level of expression of GAL4. There are promising steps in that direction with a few well-established transcription factor-binding sites dictating wing stripes and embryo segment patterns (Guss et al., 2001; Markstein et al., 2004; Moses and Rubin, 1991). With comparative analysis of the 12 sequenced *Drosophilid* genomes, the transcription factor binding site mapping projects (Gallo et al., 2006), and the antibody generation effort to map the expression pattern of transcription factors (http://www.modencode.org/), it is rational to hope that this kind of designer control element may someday exist.

An additional level of spatial expression control can be added to the UAS-GAL4 system by including **GAL80**. GAL80 is another yeast protein that binds to GAL4 and prevents it from activating transcription (Ma and Ptashne, 1987). More GAL80 may be needed to neutralize a given amount of GAL4. A GAL4 line and a GAL80 line with overlapping expression patterns can be combined (Lee and Luo, 1999). The UAS reporter line will only be expressed in places where the GAL4 is present but not the GAL80, providing a negative intersectional strategy (Suster et al., 2004). It is hard to see where a GAL80 line is expressed: there is no good antibody for immunohistochemistry and the protein is likely to be nuclear or cytoplasmic, making it difficult to extrapolate which neurons express it. It is possible to convert a GAL4 enhancer trap line into a GAL80 line by P-element replacement (Sepp and Auld, 1999) but screening for this can be hard to do visually and PCR screening is sometimes required.

This kind of intersection can also occur in time as well as space. The **TARGET** approach uses ubiquitous expression of a temperature-sensitive version of GAL80 to suppress GAL4 function while the flies are at permissive temperature (McGuire et al., 2003). The flies can be temperature-shifted, which inactivates GAL80, and now the GAL4 is able to activate reporter genes. The ramp up of GAL4 is gradual and the temperature shifting may not be appropriate for all experiments, but this strategy was a major advance for temporal as well as spatial control of gene expression. Another method, **GeneSwitch**, adds temporal control with a drug-sensitive GAL4 (Osterwalder et al., 2001; Roman et al., 2001). Animals are fed RU486, which then binds the modified GAL4 to activate gene expression. This approach requires rebuilding the GAL4 lines of interest and it also has slow kinetics (on the order of 24 h). Both methods suppress GAL4 expression during development and then allow function to be turned on; they are less effective for rapidly turning GAL4 function off. The TARGET and Gene-Switch methods have been reviewed (McGuire et al., 2004). A temperature-sensitive version of GAL4 itself is another alternative (Mondal et al., 2007). The **Tet-on/Tet-off** system requires three transgenes but allows the use of the existing GAL4 collections (Stebbins et al., 2001). It relies on modifying the reporter gene to be drug sensitive and has not been widely adopted. Modifying the reporter to be produced in a temperature-sensitive fashion produced using **inteins** might also be possible (Zeidler *et al.*, 2004). The slow kinetics of these systems is acceptable if the amount of compensation for the manipulations is minimal and if the behavior under study can be triggered acutely.

Another option for increasing the specificity of a broadly expressing GAL4 is to use it to drive an **RNAi** construct for a transcript that is only present in a subset of cells. This approach was used to identify the *fruitless*-positive median bundle neurons as the critical ones involved in some aspects of courtship behavior (Manoli and Baker, 2004). There are now several collections of RNAi lines for neural genes available (Dietzl *et al.*, 2007; Mathey-Prevot and Perrimon, 2006; Ni *et al.*, 2008; Sepp *et al.*, 2008). Screening genetically targeted RNAi lines has been used to identify the sex peptide receptor and the neurons that express it as critical components of the circuitry for female receptivity behavior (Hasemeyer *et al.*, 2009; Yapici *et al.*, 2008).

Whereas the above methods are negative intersectional strategies, in that they are used to remove part of a GAL4 pattern, positive intersectional strategies have also been developed. These allow the targeting of a reporter to areas only where two expression patterns overlap. The GAL4 protein can be split into two pieces, one of which contains the DNA-binding domain and the other of which activates transcription. (This is the basis for the yeast two-hybrid screening system.) The two pieces can be brought back together again by leucine zipper motifs with high specific affinity and reconstitute a protein that is less effective than the original GAL4 but is still able to activate transcription (Luan *et al.*, 2006b). Each half of the **split GAL4** can be expressed in different patterns, and functional GAL4 is only reconstituted in the overlap zone to drive reporter expression. This approach has been used to identify which neurons within a larger group are really responsible for driving wing expansion (***Luan et al., 2006a***). The split GAL4 technique has great potential utility but requires rebuilding the GAL4 lines of interest. Some examples of astute use of these tools for circuit bashing can be seen in ***Gao et al. (2008)*** and ***Shang et al. (2008)***.

Other positive temporal or spatial intersectional strategies involve the **FLP and Cre recombinases** (reviewed in Bischof and Basler (2008)). During mitosis, they catalyze excision and ligation of double-stranded DNA at defined DNA sequences (FRT or lox sites) (Golic, 1991; Golic and Lindquist, 1989). FLP was initially used for generating chromosomal breaks at the base of each chromosome arm using a heat-shock induced expression of flippase (Basler and Struhl, 1994; Struhl and Basler, 1993; Xu and Rubin, 1993). When the chromosomal break occurs in a dividing cell, it produces a clone of cells that are homozygous mutant in a heterozygous background. Several strategies for making targeted mosaics where specific parts of the fly (usually the eye) expressing the recombinase can become homozygous mutant. This allowed screens for genetic mutations that might have been lethal in the whole animal and was very successful at identifying components of synaptic function, for example (Blair, 2003; Newsome *et al.*, 2000; St Johnston, 2002; Stowers and Schwarz, 1999). Although recombination events are not reversible, temporal control of the initial recombination event can be achieved with a heat-shock-inducible enhancer (usually from hsp70) or a hormone-inducible motif appended to the recombinase itself (Heidmann and Lehner, 2001). A modification of this approach called mosaic analysis with a repressible cell marker (**MARCM**) combines the use of FLP with GAL4 and GAL80 to mark mutant clones within a given GAL4 pattern (Lee and Luo, 1999, 2001). The MARCM technique allows the intersection of marking based on lineage and marking based on gene expression, which represents an extremely powerful anatomical technique for visualizing cell lineages and single cells (Jefferis *et al.*, 2001).

The recombinases can be used to trigger intrachromosomal recombination events between defined sites as well. Usually this approach involves recombination to remove a stop cassette between UAS and a reporter or effector; it is sometimes called **Flp-out**. The recombination can occur in postmitotic cells and so affects a random set of cells within a given GAL4 pattern. This strategy can be used to equalize expression levels of different reporter constructs, to prolong the expression of a GAL4 that is expressed early in development, or to positively intersect two expression patterns. If the recombination is triggered in a dividing cell, this approach can be used to label neurons related by lineage as is obligatory in MARCM. For example, *TubP-FRT-STOP-FRT-GFP*, *UAS-flippase*, and *PoxN-GAL4* can be combined to cause the expression of GFP to be maintained in all the cells in which the early-expressing GAL4 was active. An enhancer trap GAL4 line could be combined with a line expressing flippase in all the glutamatergic neurons and a *UAS-FRT-STOP-FRT-GFP* to visualize only the glutamatergic neurons within the enhancer trap pattern. The recombinases are reported to work at very high efficiency, especially when catalyzing intragenic—rather than interchromosomal—recombination, and there is a range of matched recombinase binding sites that work in *Drosophila* (Heidmann and Lehner, 2001; Horn and Handler, 2005; Oberstein *et al.*, 2005; Rodin and Georgiev, 2005; Siegal and Hartl, 1996).

To subdivide a GAL4 pattern for imaging, a clonal approach like MARCM or a random approach using a recombinase removes a stop cassette in postmitotic cells are effective alternatives (Chiang *et al.*, 2004; Marin *et al.*, 2002; Wong *et al.*, 2002). These approaches can be used for behavioral analysis, but large numbers of individual animals are needed to get statistical confidence that particular neurons really correlate with a given behavioral defect (Gordon and Scott, 2009a; Kimura *et al.*, 2008; Shang *et al.*, 2008; Yang *et al.*, 2009).

Recently, alternative two-component systems have been transported to the fly. The *yeast* **LexA** transcription factor and the lexOp DNA sequence to which it binds appear to work in flies as well. This allows independent targeting of different transgenes (Lai and Lee, 2006). For example, one might target UAS-mCD8-GFP to the presynaptic side of the neuromuscular junction with VGlut-GAL4 and lexOP-CD2-mRFP to the postsynaptic side with an MHC-LexA. LexA and GAL4 can also be combined to expand the repertoire of intersectional strategies. Other transcription factor—binding site systems are under development.

   With the library of expression patterns that can be generated by the GAL4-based strategies described above and summarized in Fig. 3.2, we have the tools to image and manipulate neural circuits with unprecedented spatial and temporal precision.

III. IMAGING NEURONS

To visualize the neurons in which GAL4 is expressed, the membrane-targeted green fluorescent protein encoded by **UAS-mCD8-GFP** is most commonly used (Lee and Luo, 1999); alternative anatomical reporters are listed in Table 3.1. The endogenous or intrinsic fluorescence of GFP in live or briefly fixed tissue is usually sufficient to detect the small processes of neurons, but when the tissue can be fixed, the signal is often amplified with primary antibodies against CD8 or GFP itself and bright, photostable, dye-coupled secondary antibodies. For an example protocol, see Wu and Luo (2006). This also allows counterstaining with the mouse nc82 monoclonal antibody to label the whole synaptic neuropil and provides a broad landmark for registering different preparations to a common standard (Jenett et al., 2006; Rein et al., 2002).

   While UAS-mCD8-GFP provides a good staining of neuronal processes for anatomical analyses, the cytoplasmic **UAS-eGFP** has been reported to be the most innocuous for electrophysiology (Su and O'Dowd, 2003); here the endogenous brightness is essential since GFP here is used to target electrodes in live preparations. GFP has been optimized for brightness, photostability, and pH insensitivity. Most of the GFP in current use is codon optimized for vertebrates, rather than the original jellyfish, and contains the S65T point mutation; thus, it should more precisely be called EGFP (Yang et al., 1996).

   The choice of alternative colors of fluorescent proteins is dizzying (Giepmans et al., 2006; Shaner et al., 2005). It is possible to image different neural populations in different colors using direct enhancer fusions or orthogonal expression systems (GAL4 and LexA). We now have photoactivatable and photoswitchable fluorophores, and fluorophores that change color over time (Terskikh et al., 2000) are reviewed (Lippincott-Schwartz and Patterson, 2008). Timer was used to show that the inner fibers of the mushroom bodies are younger than the outer fibers, indicating that unlike tree rings, the

Table 3.1. UAS-Reporters for Visualizing Neurons

Construct name	Localization	Comments	References
Anatomy			
UAS-mCD8-GFP	Membrane	Can also be detected with antibodies to CD8 or GFP	Lee and Luo (1996)
UAS-myr-mRFP	Membrane	Uses endogenous fluorescence of mRFP	H. Chang, flybase
UAS-eGFP	Cytoplasm	Electrophysiologically neutral	Su and O'Dowd (2003)
UAS-nls-GFP	Nucleus	Both GFP and lacZ fusions exist	Robertson et al. (2003)
UAS-nSyb-GFP UAS-Syt-GFP UAS-Syt-HA	Synapses	Visualized with antibody to HA	Estes et al. (2000), Robinson et al. (2002), and Zhang et al. (2002)
UAS-DsCam17.1-GFP	Dendrites	May change dendrite morphology	Wang et al. (2004a)
UAS-Rdl-HA	Postsynapse	Visualized with antibody to HA	Sanchez-Soriano et al. (2005)
UAS-cac-GFP	Active zones	Tested at neuromuscular junction	Kawasaki et al. (2004)
UAS-tau-lacZ	Axons	Both GFP and lacZ fusions exist; may affect neuron health	Callahan and Thomas (1994) and Hidalgo et al. (1995)
UAS-nod-lacZ	Dendrites	Both GFP and lacZ fusions exist	Anderson et al. (2005) andClark et al. (1997)
UAS-GAP-GFP	Axons	Tested at neuromuscular junction	Ritzenthaler et al. (2000)
UAS-PA-GFP	Cytoplasm	Activated by 710 nm light	Datta et al. (2008)
UAS-Timer	Cytoplasm	Switched from red to green over several hours	Verkhusha et al. (2001)
Activity			
UAS-GCaMP	Cytoplasmic	Calcium sensor; various improved versions exist	Wang et al. (2003a,b)
UAS-Cameleon	Cytoplasmic	FRET calcium sensor; Synapcam is a synaptically targeted variant	Fiala et al. (2003), Guerrero et al. (2005), and Hendel et al. (2008)
UAS-Camgaroo	Cytoplasmic	Calcium sensor	Yu et al. (2003)
UAS-TN-XXL	Cytoplasmic	Calcium sensor using troponin C	Mank et al. (2008)
UAS-D3cpv	Cytoplasmic	Redesigned M13 peptide	Hendel et al. (2008)
UAS-GFP-Aequorin	Cytoplasmic	Bioluminescent Ca^{2+} indicator	Martin et al. (2007)
UAS-FlaSh	Membrane; synapse	Voltage sensor; based on a pore-mutated Shaker voltage-gated K+ channel subunit	Siegal and Isacoff (1997)
UAS-hVos	Membrane	Hybrid voltage sensor	Sjulson and Miesenbock (2008)
UAS-SynaptopHluorin	Synaptic vesicles	Vesicle release detector	Miesenbock et al. (1998)
UAS-Epac1-camps	Cytoplasmic	cAMP level reporter	Shafer et al. (2008)

late-growing axons actually push up through a bundle of established tracts (Verkhusha *et al.*, 2001). Photoactivatable GFP has been used to trace a group of axons with particular odor response profiles (***Datta et al., 2008***).

There are options for targeting reporter proteins to different subcellular compartments. One might use the dendritic and synaptic reporters to deduce input and output zones in a given GAL4 pattern to hypothesize about connectivity or information flow. To visualize neural processes, fusions to the tau motor protein were initially popular and they provide excellent labeling (*UAS-tau-lacZ* and **UAS-tau-GFP**) (Callahan and Thomas, 1994; Hidalgo *et al.*, 1995), but they are deleterious to many neural types (Williams *et al.*, 2000). *UAS-GAP-GFP* (Ritzenthaler *et al.*, 2000) also labels axons. The T-cell membrane-targeting motifs from CD2, CD4, and CD8 (mouse or rat) and myristylation sequences from c-src seem to bring fluorescent proteins to the membrane efficiently in insect cells. To visualize nuclei, nuclear localization signals work well: *UAS-nls-lacZ* and **UAS-nls-GFP** (Hiromi, unpublished Bloomington stock #3955; Robertson *et al.*, 2003). Synaptic targeting can be achieved with fusions to SNARE proteins nSyb and synaptotagmin: **UAS-nSyb-GFP, UAS-Syt-GFP**, *UAS-Syt-HA*, and *UAS-nSyb-mRed* (Estes *et al.*, 2000; Raghu *et al.*, 2007; Robinson *et al.*, 2002; Zhang *et al.*, 2002). There is an active zone marker **UAS-cac-EGFP** that works at the neuromuscular junction in high copy number (Kawasaki *et al.*, 2004). Labeling dendrites or postsynaptic densities is currently the most problematic, but there are reports that it can be done with **UAS-dsCam17.1-GFP** (cell adhesion molecule: Wang *et al.*, 2004a) or **UAS-Rdl-HA** (ionotropic GABA receptor: Sanchez-Soriano *et al.*, 2005). *UAS-nod-GFP* (Andersen *et al.*, 2005; Clark *et al.*, 1997), a fusion to another minus-end directed microtubule motor protein, also labels dendrites in some cell types. To move from the possibility of connections suggested by proximity of axons and dendrites to actual connectivity is an important leap that requires further evidence.

The **GRASP** technique for confirming that two neurons are synaptically connected by separately targeting expression of halves of GFP to the pre- and postsynaptic sides of a synapse to reconstitute functional fluorescence (Feinberg *et al.*, 2008), has now been adapted for the fly (Gordon and Scott, 2009a,b). An activity-dependent trans-synaptic tracer that works in many types of neurons would be extremely beneficial for exploratory investigation of neural connectivity, but in spite of hard work in many labs, none is currently available. Electron microscopy can show the presence of synapses and specific neuron classes can be targeted using the GAL4 system to drive *UAS-CD2-HRP* (Larsen *et al.*, 2003); the reaction product of this extracellularly tethered horse radish peroxidase is electron-dense. Synaptic specializations and vesicles may be visible. The number of synaptic contacts and the quantity of docked vesicles might provide some indication of the strength of the connection but the excitatory or inhibitory nature must be deduced by other means.

Please keep in mind some caveats. It is not certain that all of the manipulations we do to visualize neurons are neutral. High levels of GAL4 or reporter proteins may be toxic or alter cell morphology (Kramer and Staveley, 2003). Membrane-targeted proteins may be expressed highly enough to disrupt normal membrane properties. It is possible to have pre- and postsynaptic contacts on the same neurite (Olsen and Wilson, 2008b; Raghu et al., 2007), which makes the analysis of circuitry at the light level more challenging. Confocal microscopy is typically used to visualize these reporters and optimal tissue clearing, laser ramping, and data collection standards are not always achieved. Some serious pitfalls are astutely enumerated in *Ito et al. (2003)*. The level of the visible reporters may not match the level of effectors expressed, making it difficult to draw firm conclusions about which neurons visualized by *UAS-mCD8-GFP* are the ones responsible for the behavior seen with *UAS-Shibire*[ts1] expressed by the same GAL4 line. Inserting all reporters and effectors into defined loci with the integrase system may help here by eliminating position effect varigation, and tagging the effectors directly with epitope tags or coexpressing reporters and effectors together with an internal ribosome entry site (IRES) or 2A self-cleaving peptide (Trichas et al., 2008) may go some way toward ameliorating these concerns, but interpretations should be cautious. Detection thresholds for staining and behavior may be very different.

IV. FUNCTIONAL IMAGING: WATCHING NEURONAL ACTIVITY

The promise of functional neuroimaging is to be able to see activity in the processes or compartments of a single identified neuron, or to assay activity in several identified neurons at once, to watch circuit computations in action. Functional neuroimaging can be used to identify relevant neurons or to investigate exactly what previously identified neurons are doing during behavior performance. For neuronal activity, one can monitor membrane voltage or changes in calcium concentration; these reporters have been developed primarily in vertebrate systems and are discussed in greater detail below. The versions of these reporters that are available in the fly are listed in Table 3.1. Reporters for other biological activities exist that have relevance for neural function. There are new reporters for glutamate, usually an excitatory neurotransmitter (Hires et al., 2008b), cAMP levels (Shafer et al., 2008), Creb (Belvin et al., 1999), receptor activation (Barnea et al., 2008), and some kinase activities (Burrone, 2005; Tsien, 2005). **UAS-synaptopHluorin**, pH-sensitive fluorescent protein coupled to neural-synaptobrevin can be used to visualize synaptic vesicle release (Miesenbock et al., 1998; Ng et al., 2002). SynaptopHluorin has also been used to show that additional neurons become active during the establishment of an olfactory memory (Yu et al., 2004, 2005). Optical reporters represent a powerful, relatively noninvasive, technique for investigating neural circuits.

For neural circuit mapping, reporters that act over a longer timescale may be useful if they help identify brain regions that are active when a behavior is performed repeatedly. In mice, there have been attempts to harness the **immediate early genes** whose transcription is up-regulated by neural activity for this purpose (Barth et al., 2004; Mongeau et al., 2003; Reijmers et al., 2007; Wang et al., 2006). Exactly what these changes in gene expression mean is subject to intense debate. So far, attempts to transport this technique to flies have not been reported.

A. Voltage sensors

Just as it is appealing to be able to activate neurons in a way that mimics action potentials, it would be terrific to see neural activity at the resolution of action potentials. For an example of what can be done with really precise voltage measurements in multiple neurons simultaneously, see the work identifying the neurons that best correlate with the decision to swim rather than crawl in the leech (Briggman et al., 2005). While most of the neurons that drive swimming and crawling are part of a shared network, a few correlate with only one of the two behaviors (Briggman and Kristan, 2006). The fast kinetics from voltage-sensitive dyes are powerful, but the dyes cannot be specifically targeted, penetrate different tissues unevenly, and provide poor spatial resolution. It would be ideal to have a genetically encoded voltage sensor; the current state of the field is well reviewed in Baker et al. (2008) and summarized below.

The original voltage-sensing fluorescent proteins were based on ion channels. **FlaSh** tethered GFP to a pore-mutant version of the *Drosophila* Shaker potassium channel close to the membrane so that movement of the voltage-sensing helix affected the fluorescence (Siegel and Isacoff, 1997). This produced a change in fluorescence in *Xenopus* oocytes (5% change with an -80 mV depolarization) but was not able to detect voltage changes in neurons and had kinetics too slow to resolve individual action potentials (on: 100 ms; off: 60 ms). Optimization of the fluorophore improved the response time to $\sim$5 ms but did not make significant improvement in the amount of fluorescence change or the usability in neurons (Guerrero et al., 2002). An alternative to FlaSh, voltage-sensing fluorescent protein 1 (**VSFP1**) was FRET based and used the isolated voltage-sensing S4 domain of the vertebrate potassium channel Kv2.1 (Sakai et al., 2001). Sodium channel protein-based activity reporting construct (**SPARC**) fused GFP between the first and second 6 transmembrane repeat domains of the voltage-sensitive sodium channel rSkM1 (Ataka and Pieribone, 2002). All three of these channel-based voltage-sensors performed poorly in neurons because they failed to localize well to the plasma membranes, resulting in low signal and high noise levels from the mislocalized fluorescence (Baker *et al.*, 2007a). Endogenous ion channel levels in the plasma membrane are tightly regulated to tune neural activity; perhaps the engineered voltage sensors based on ion channels are subject to the same regulatory mechanisms.

New voltage sensors under development use protein domains from enzymes rather that ion channels (Murata *et al.*, 2005; Ramsey *et al.*, 2006; Sasaki *et al.*, 2006; Tombola *et al.*, 2008). There is some hope that these will ameliorate the plasma localization and protein density limitations that plague the channel-based constructs. This may increase the detectable change in fluorescence. Additional improvements occur all the time (Tsutsui *et al.*, 2008; Villalba-Galea *et al.*, 2009), but whether the sensors will be able to follow the speed of action potentials in neurons *in vivo* is still uncertain.

Since voltage-sensitive chemical dyes can provide the high signal-to-noise ratio and fast kinetics desired for a voltage sensor with action potential resolution, there was some excitement about hybrid systems that couple a dye to a genetically encoded fluorescent donor or acceptor protein (Chanda *et al.*, 2005) which could provide the spatial localization the dyes alone lack. Unfortunately the hybrid voltage sensor (**hVOS**) approach that has been best tested in flies has not performed as well as hoped (Sjulson and Miesenbock, 2008). In a combination of modeling calculations and experiments where flies expressing a membrane-tethered GFP as a FRET donor were exposed to dipicrylamine (DPA, a voltage-sensitive FRET acceptor dye), Sjulson and Miesenbock showed that to see a significant fluorescence change even in a large group of neurons firing synchronously, such a high concentration of dye was required that the quantity of dye intercalating in the membrane changed its capacitance sufficiently to stifle action potentials. Other variants of the dye/genetic hybrid approach are possible but have yet to show positive results for voltage sensing in neurons (Hinner *et al.*, 2006; Lavis *et al.*, 2006).

B. Genetically encoded calcium indicators

More widely used than voltage sensors, calcium sensors act as a proxy to report neuronal activity. Calcium dynamics within neurons are complicated (Yasuda *et al.*, 2004). When a neuron fires an action potential, its membrane depolarizes in a propagating wave moving along the axon toward the synaptic terminal. This depolarization triggers the opening of voltage-gated Ca^{2+} channels (encoded by *cacophony* in *Drosophila*: Kawasaki *et al.*, 2000). The local influx of Ca^{2+} triggers the fusion of vesicles containing neurotransmitter with the plasma membrane, causing the neuron to pass information on to its postsynaptic partners. Repeated action potentials increase the local Ca^{2+} concentration in the neurons and thus Ca^{2+} levels are an indicator of how active the neuron is. Genetically encoded calcium indicators (GECIs) can also be used to look at the Ca^{2+} dynamics in subcellular compartments such as dendritic branches, where calcium enters through nonselective ion channels, including the ligand-gated ionotropic neurotransmitter receptors such as the glutamate and voltage-gated NMDA receptor. *Drosophila* has NMDA receptors (Xia *et al.*, 2005) but whether their contribution to Ca^{2+} influx can be seen with GECIs has not been explored. Pumping the Ca^{2+} into intracellular stores in the endoplasmic reticulum or out of the cell with Ca^{2+}-ATPase pumps (PMCA) gradually restores the Ca^{2+} levels.

There are highly sensitive chemical indicators of calcium level (Fura dyes and Calcium green; for example use in fly, see Wang *et al.* (2001)), and these can be used with genetic markers of cell identity (Ritter *et al.*, 2001; Yaksi and Friedrich, 2006). There are also a variety of GECIs (reviewed in Hires *et al.* (2008a) and Miyawaki *et al.* (2005)). These are composed of a fluorescent protein (or two) and a peptide that changes conformation upon Ca^{2+} binding (calmodulin or troponin C). **Camgaroo** is a circularly permuted GFP with the calmodulin Ca^{2+}-binding domain at one end and the M13 calmodulin-binding peptide (from myosin light chain kinase) at the other; it undergoes a reversible conformational change upon ion binding that increases the fluorescence of GFP (Baird *et al.*, 1999). Camgaroo was used in *Drosophila* to visualize activity in the mushroom bodies in response to exogenously applied acetylcholine (Yu *et al.*, 2003). **Pericams** (Nagai *et al.*, 2001) and **GCaMPs** (Nakai *et al.*, 2001) use a similar strategy to detect an increase in Ca^{2+}. The GCaMP sensors are currently the most highly developed of these. Although membrane targeting GCaMP2 does not improve its performance (Mao *et al.*, 2008), new variants have the ability to reliably detect short trains of action potentials in some cell types and more improvements are expected soon. GCaMP and its derivatives have been used to map where different types of tastes and odors are processed (Fischler *et al.*, 2007; Marella *et al.*, 2006; Suh *et al.*, 2004; Wang *et al.*, 2003a) and to detect neuronal activity in the mushroom bodies during olfactory conditioning (Yu *et al.*, 2006). It may also be possible to use this type of imaging to identify which neurons within a complicated GAL4 pattern have activity correlated with the behavior under study and thus narrow down complicated expression patterns to spot the relevant neurons (but see caveats below).

The **cameleon** sensors also use calmodulin and the M13 peptide but in this case Ca^{2+} binding brings together two different fluorophores for fluorescence resonance energy transfer (FRET) (Miyawaki *et al.*, 1997, 1999). Ratiometric imaging of this type has been particularly helpful to compensate for movement artifacts (Kerr *et al.*, 2000). Recent variants have optimized the choice of fluorescent donor–acceptor pairs to maximize FRET and reduce interference with endogenous Ca^{2+} sensors (Yellow cameleons and D3cpv: Nagai *et al.* (2004) and Palmer *et al.* (2006)). Fiala *et al.* used Cameleon in *Drosophila* to examine olfactory responses in projection neurons and to demonstrate that dopaminergic neurons fire strongly in response to electrical shock during olfactory conditioning assays (Fiala *et al.*, 2002; Riemensperger *et al.*, 2005). Cameleon has also been useful specifically for mapping novel neural circuits: Liu *et al.* used the reporter to identify the thermosensing neurons in larva (**Liu *et al.*, 2003**). Synapcam is a synaptically targeted version of cameleon (Guerrero *et al.*, 2005) that shows that more distal boutons along a larval neuromuscular junction have higher levels of Ca^{2+} influx, a result that agrees with Ca^{2+} sensitive dye experiments (Lnenicka *et al.*, 2006).

TN-XXL is an alternative FRET-based Ca^{2+} sensor. Instead of calmodulin and the M13 peptide, it exploits a similar domain from troponin C (which is not present in neurons), and so may not interfere with endogenous calmodulin function. It can be activated by the longer wavelengths required for two-photon imaging *in vivo* in flies and mice. It has reasonable fluorescence change signal and may perform better for detecting changes when the overall Ca^{2+} concentration is low. Its performance has been characterized (Mank *et al.*, 2006, 2008).

As an alternative to fluorescence, a few groups have used GFP-Aequorin constructs to measure Ca^{2+} changes with bioluminescence (Martin *et al.*, 2007; Rosay *et al.*, 2001). This sensor requires coelenterazine as a cofactor. For very long timescale experiments, this is a possible alternative sensor.

There are problems with all of the GECIs. They tend to have small dynamic range, poor sensitivity, and slow kinetics. Calcium is an indirect proxy for neural activity and the indicators distort the kinetics of the calcium signal. Several recent reviews have compared the available Ca^{2+} indicators (Hendel *et al.*, 2008; Mao *et al.*, 2008; Martin, 2008; Miesenbock and Kevrekidis, 2005; Pologruto *et al.*, 2004; Reiff *et al.*, 2005). The best choice may depend on the exact preparation and expected Ca^{2+} concentration range. In the best cases it may be possible to detect single action potentials with reasonable reliability, but this has not yet been done in the fly. If the action potentials are sparse, the rise time of the Ca^{2+} indicators is sufficient to detect them with high reliability; the decay time is slower, so if the action potentials occur too close together, they cannot be individually resolved, but rate can be estimated by deconvolution (Kerr and Denk, 2008; Wallace *et al.*, 2008). In many neurons multiple spikes are required to generate a visible fluorescence change and the temporal precision of the indicators may make this difficult. The calcium indicators may buffer the Ca^{2+} they detect and may interfere with normal Ca^{2+} binding proteins. They are not able to detect subthreshold or graded changes in membrane potential. The Ca^{2+} signal almost always under-represents the number of active neurons involved because of the high thresholds of activity required to trigger the sensors. In any case, careful interpretation and system-specific validation is needed to determine exactly what the detected change in Ca^{2+} concentration represents—and what it may miss (Hendel *et al.*, 2008; Jayaraman and Laurent, 2007).

GECIs are a powerful way to identify the neurons involved in particular behaviors or circuits, but can sometime yield different results than electrophysiology. Ca^{2+} dynamics measured with GECIs are slower than changes in membrane potential; this allows summation of weak signals but makes it hard to resolve fast spike trains. Several research groups investigated the transformation of information that occurs at different relay points in the olfactory circuit. The results obtained with GCaMP and SynaptopHluorin differed from that obtained with electrophysiological recordings (Ng *et al.*, 2002; Olsen and Wilson, 2008b; Root *et al.*, 2007; Shang *et al.*, 2007; Wang *et al.*, 2003a; Wilson *et al.*, 2004). Since it is now possible—albeit difficult—to record from neurons in the fly brain during sensory experience (Wilson *et al.*, 2004), it is possible to better calibrate the genetic reagents we use to inhibit, activate, and monitor neurons (Jayaraman and Laurent, 2007).

There is a long history of **electrophysiological recording** from neurons, muscles, and sensory structures *Drosophila*. Technical reviews include (Broadie, 2000a,b; Matthies and Broadie, 2003). Electrophysiological methods have been critical for assaying ion channel properties, synaptic vesicle release and recycling machinery, neurotransmitter identity, mechanisms of synaptic plasticity, and sensory information coding. The new genetic tools for manipulating neural activity (discussed below) have been tested by electrophysiology. For circuit analysis in particular, electrophysiological techniques have been instrumental in identifying brain regions involved in specific behaviors, establishing the temporal code of action potentials generated in response to sensory stimuli, and demonstrating connectivity by paired recording or in combination with activation by Channelrhodopsin or imaging with GCaMP. Although electrophysiology in the fly is limited to one—or at most a few—neurons at a time, it provides unparalleled sensitivity and temporal precision for monitoring neural activity. The technical challenges of recording from small, deep brain neurons in a behaving animal should not be underestimated. Table 3.2 lists some of these electrophysiological techniques and example papers where they are used.

V. CONTROL OF NEURAL ACTIVITY

There are many strategies for manipulating neurons once one has a reproducible way to target them. There are cell killers based on toxins or genes that promote programmed cell death; ion channels and proteins that interfere with a neuron's excitability; toxins, and mutations that disrupt the synaptic vesicle cycle; and a slew of enzyme-specific blockers. I refer to these UAS constructs collectively as "effectors" rather than "reporters," which are usually fluorescent ways to visualize cells. All of these effectors have pros and cons associated with them; available reagents are summarized in Table 3.3 and discussed below.

Table 3.2. Electrophysiological Techniques in *Drosophila*

Technique		References
Culture		
Embryonic	Neuroblasts from gastrulating embryos are isolated, dissociated, and induced to extend neurites in culture; whole cell patch recordings are performed	O'Dowd (1995), O'Dowd and Aldrich (1988), and Seecof et al. (1971)
Giant neurons	Neuroblasts are harvested from embryos and then the last cell divisions are blocked to create large multinucleate neurons that can be targeted with electrodes for whole cell patch recording	Saito and Wu (1991)
CNS neurons	Cells are cultivated from embryos and larvae and genetically labeled neurons are targeted for whole cell patch clamp recording to study electrical properties of the neurons	Sicaeros et al. (2007) and Wright and Zhong (1995)
Photoreceptors	Adult or pupal ommatidia are cultured for subsequent whole cell patch clamp recording to characterize electrical properties in genetically identified neurons	Hardie (1991)
Neuromuscular junction (NMJ)		
Giant fiber	Flies are immobilized and recordings from motoneuron, muscle and/or the giant fiber axon is preformed. The giant fiber is electrically stimulated through tungsten electrodes placed in the eyes or brain	Elkins and Ganetzky (1990), Engel and Wu (1996), Fayyazuddin et al. (2006), Koenig and Ikeda (1983), and Tanouye and Wyman (1980)
Larval NMJ	Larvae are filleted out and intracellular voltage recordings from the muscle can measure both evoked junctional potentials (EJPs) and excitatory junctional currents (EJCs). Two electrode voltage clamp (TEVC) recordings from the muscle have been used to identify membrane currents	Imlach and McCabe (2009), Jan and Jan (1976), Singh and Wu (1989), Wu and Haugland (1985), and Zhong and Wu (1991)
Larval motor nerves	Recording and stimulating from different points along the nerve bundle shows conduction defects and direction of action potential propagation	Wu et al. (1978)
Embryonic NMJ	Whole cell patch clamp and perforated patch recordings from developing muscle are possible in dissected young embryos (<17 h AEL). Older embryos require dissection at 16 h AEL and culturing to the appropriate developmental stage	Broadie and Bate (1993)

(*Continues*)

Julie H. Simpson

Table 3.2. (*Continued*)

	Technique	References
Embryonic and larval motor neurons	In the filleted animal, whole cell recordings from identified motoneurons, as well as loose patch recordings over synaptic boutons, are also possible	Baines and Bate (1998), Baines et al. (2006), Choi et al. (2004), and Rohrbough and Broadie (2002)

Sensory periphery

Photoreceptors	Flies are immobilized and a small hole made in their cornea to allow *in vivo* recordings using sharp glass microelectrodes. This allows study of signal processing and response dynamics of photoreceptors	Juusola and Hardie (2001) and Niven et al. (2003)
Large monopolar cells (lamina)	Small corneal openings in an immobilized fly's eye allow sharp glass microelectrode recordings. Information processing at first synapse of the system can be studied. LMCs are identified by their distinctive electrical properties	Zheng et al. (2006)
Electroretinograms	Extracellular recording measures light-induced depolarization of photoreceptors and synaptic activation of second order neurons	Alawi and Pak (1971), Hotta et al. (1969), and Kelly and Suzuki (1974)
Mechanosensory bristles	Extracellular transepithelial potential recording measures neuronal response to bristle deflection	Dickinson and Paulka (1987) and Kernan et al. (1994)
Electroantennograms	Extracellular recordings from the antennal nerve measures gross output of olfactory sensory neurons	Borst (1984) and Venard and Pichon (1981, 1984)
Olfactory receptor neurons	Flies are immobilized for extracellular recordings using low-impedance glass or tungsten electrodes; recordings are made from base of olfactory sensilla in antennae and maxillary palp which allows isolation of activity of single olfactory receptor neurons in response to odors	Clyne et al. (1997), de Bruyne et al. (1999), Hallem et al. (2006), and Kreher et al. (2008)

Central nervous system (CNS)

Mushroom body kenyon cells and circadian pacemaker neurons from isolated whole brain explants	Whole brains are isolated from adult flies and prepared for whole cell patch clamp from genetically labeled neurons using differential interference contrast (DIC) imaging; this permits examination of electrical properties of neurons under different conditions (e.g., during sleep/awake phases in circadian cycle)	Cao and Nitabach (2008), Gu and O'Dowd (2006, 2007), and Sheeba et al. (2008)

(*Continues*)

Table 3.2. (*Continued*)

	Technique	References
Antennal lobe projection neurons from isolated whole brain explants	Whole brains and antennae are isolated from adult *Drosophila* and bathed in saline for loose patch recordings from genetically labeled neurons targeted using two-photon imaging; recordings allow detection of action potentials in response to odor	Root *et al.* (2007)
Antennal lobe projection neurons and interneurons	Flies are immobilized and dorsal sections of cuticle, trachea and sheath removed to expose antennal lobes; the brain is bathed in saline keeping antennae in air for *in vivo* whole cell patch clamp and loose patch recordings from genetically labeled neurons in response to odors (performed under visual guidance using DIC optics or two-photon imaging)	Bhandawat *et al.* (2007), Datta *et al.* (2008), Jayaraman and Laurent (2007), Olsen and Wilson (2008a,b), Wilson and Laurent (2005), and Wilson *et al.* (2004)
Mushroom body kenyon cells	Flies are immobilized and posterior sections of cuticle, trachea and sheath are removed to expose mushroom body. The brain is bathed in saline for *in vivo* whole cell patch clamp recordings from genetically labeled neurons in response to air-delivered odors; recordings are performed under visual guidance (DIC)	Murthy *et al.* (2008) and Turner *et al.* (2008)
Lobular plate interneurons	Flies are immobilized and lateral-posterior sections of cuticle, trachea and sheath removed to expose lobula plate. The brain is bathed in saline while eyes remain in air. This allows visually guided whole cell patch clamp recordings of vertical-sensitive neurons of the lobula plate tangential system in response to visual patterns	Joesch *et al.* (2008)

A. Cell killers

One way to genetically ablate neurons is to express toxins that disrupt protein synthesis. Two such toxins are the poetically named "Blue Death" (**UAS-diptheria toxin A** from bacteria (Lin *et al.*, 1995)) and **UAS-ricinA** (from castor beans (Hidalgo and Brand, 1997; Hidalgo *et al.*, 1995)). Cold-sensitive versions of Ricin exist, which adds a measure of temporal control (Allen *et al.*, 2002; Moffat *et al.*, 1992). Expression of proapoptotic genes like **UAS-reaper, grim**, and **hid** can also be used to induce cell death (Zhou *et al.*, 1997).

Table 3.3. UAS-Effector Constructs for Manipulating Neural Activity

	Encodes	Function	Cell type effected	Inducible or reversible?	Comments	References
A. Cell killers						
Diptheria toxin A	Toxic polypeptide from bacteria	Protein synthesis inhibitor	All cells (neurons and nonneurons)	No	Weaker version: DTI–attenuated mutant I; can take hours for cell death to occur	Lin et al. (1995)
RicinA	Toxic polypeptide from castor bean	Protein synthesis inhibitor	All cells	Cold-sensitive version is inducible	Temperature-sensitive version; can take hours for cell death to occur	Hidalgo and Brand (1997) and Moffat et al. (1992)
Reaper, grim, hid	Proapoptotic genes from Drosophila	Induce apoptosis via caspases	All cells	No	Can take hours for cell death to occur; some genes work better in combination depending on cell type	Zhou et al. (1997)
B. Inhibitors						
Tetanus toxin (TNT or TeTxLC)	Toxic light chain from bacteria	Cleaves syb/VAMP and blocks vesicle fusion	Neurons (chemical transmission due to small SVs)	No	May not be effective in all neurons (see Thum et al. 2006); unclear how effective TNT blocks DCV release	Martin et al. (2002) and Sweeney et al. (1995)
Shibirets1	Dominant/negative mutant dynamin gene from Drosophila	Blocks endocytosis	Neurons (chemical transmission)	Temperature-inducible and rapidly reversible	May effect not just endocytosis, but also other vesicle mobilization properties; may also effect nonneuronal cells	Kitamoto (2001)

Kir2.1	Vertebrate inwardly rectifying potassium channel	Open at rest—decreases excitability of cell by hyperpolarization	Neurons (chemical and electrical transmission), muscle	No	Channels can be blocked with barium	Baines et al. (2001), Paradis et al. (2001), and Wu et al. (2008)
dOrk-deltaC	Modified Drosophila open rectifier potassium channel	Constitutively open—decreases excitability of cell by hyperpolarization	Neurons (chemical and electrical transmission), muscle	No		Nitabach et al. (2002)
EKO	Modified Drosophila voltage-sensitive potassium channel	Open at rest—decreases excitability of cell by hyperpolarization	Neurons (chemical and electrical transmission), muscle	No	Channels can be blocked by 4-AP	White et al. (2001)
Halorhodopsin (NpHR)	Chloride pump from halobacteria	Opens in response to yellow/green light and hyperpolarizes cell	Neurons (chemical and electrical transmission), muscle	Light inducible, reversible	Requires retinal cofactor	Unpublished
C. Activators NaChBac	Bacterial voltage gated sodium channel	Opens at −60 mV—increases excitability of cell by making it easier to depolarize	Neurons (chemical and electrical transmission), muscle	No	Has been seen to deplete neurohormone of neurohormone, which renders the cell effectively inactive in older animals; also has been shown to decrease firing frequency in some neurons while increasing the AP size	Nitabach et al. (2006) and Sheeba et al. (2008)

(Continues)

Table 3.3. (Continued)

	Encodes	Function	Cell type effected	Inducible or reversible?	Comments	References
TrpVR1, TrpA1	Transient receptor potential cation channel	Opens in response to various physical/chemical stimuli	Neurons (chemical and electrical transmission), muscle	Induced by capsaicin, acid, >43 °C heat; reversible	TrpA1/high temperature; TrpM8/cold temperature; TrpV3/warm temperature; TrpVR1 has been used in combination with a caged capsaicin for light gating	Marella et al. (2006) and Rosenzweig et al. (2008)
Eag-DN, Shaker–DN, Shaw–DN	Dominant/negative voltage-gated K channel	Increases excitability of cell by depolarization of membrane and preventing repolarization after AP	Neurons, muscle (only cells that normally expressing these channels)	No	May only work in cells normally expressing these channels	Broughton and Greenspan (2004), Hodge et al. (2005), and Mosca et al. (2005)
P2X2	Ionotropic purinoceptor	Opens in response to light, depolarizes cell	Neurons, muscle	Induced by light (caged ATP); reversible	Requires caged ATP as ligand	Lima and Miesenbock (2005)
Channelrhodopsin (ChR2)	Cation channel from algae	Opens in response to light, depolarizes cell	Neurons, muscle	Induced by blue light, with retinal cofactor; reversible	Volvox varient/red-shifted light; in fly, requires retinal cofactor	Hwang et al. (2007), Schroll et al. (2006), and Suh et al. (2007)

These techniques are compared in *Drosophila* Protocols (Sullivan *et al.*, 2000). It should also be possible to laser ablate cells as is done in *Caenorhabditis elegans* or kill them by expressing phototoxic proteins (Bulina *et al.*, 2006), although neither technique is common in flies. Many of the reagents that kill cells work better during development. A killer expressed only in the adult may not kill the cell or may take hours to act; it is a good idea to include a fluorescent reporter with the cell killer so that one can be sure the cell is really gone. If cell death does occur early in development, there is the possibility that the fly will be able to compensate for the cell loss and that cells that normally play a key role in circuits will be missed because alternative circuits are being used. In addition to the cell's function, it may serve as a scaffold for the growth and path finding of other neurons, so there is no guarantee that a phenotype from the loss of a given cell is truly cell-autonomous. These reagents kill all cell types, not just neurons.

As opposed to killing a cell outright, one can block its function in a variety of ways. It is possible to express constitutively active or dominant-negative versions of enzymes, motor proteins, and transcription factors critical for synaptic plasticity or neural function. There are active and inactive versions of CamKII (Griffith *et al.*, 1993; Koh *et al.*, 1999), protein kinase A (PKA) (Kiger *et al.*, 1999; Li *et al.*, 1995), heterotrimeric G proteins (Connolly *et al.*, 1996; Ferris *et al.*, 2006), CREB (Perazzona *et al.*, 2004; Yin *et al.*, 1994, 1995), fos and jun (Eresh *et al.*, 1997), and glued (Allen *et al.*, 1999). These manipulations are also not neuron specific and have slow time courses of activity but have been productively used to identify neurons responsible for particular behaviors. For examples, see Rodan *et al.* (2002), which used UAS-PKA^{inh} for the mapping of brain regions involved in ethanol response and see Joiner and Griffith (1999), which used the CamKII inhibitor *UAS-ala* for mapping circuits needed for courtship conditioning. All of these reagents affect many cell types, not just neurons. In some cases it is desirable to restrict the action of the effector to neurons; some options for this are described below.

B. Synaptic vesicle blockers

The *Clostridium* bacteria produce some of the most potent neurotoxins known: tetanus toxin and botulinum toxin. Each toxin is composed of a heavy and a light protein chain. The heavy chain controls their membrane binding and intracellular trafficking while the light chain encodes a protease that cleaves SNARE components of vesicle release machinery (Lalli *et al.*, 2003; Schiavo *et al.*, 2000). **UAS-TNT** (also known as *UAS-TeTxLC*) expresses the light chain of tetanus toxin and cleaves neural synaptobrevin (VAMP), making it a powerful reagent to specifically block vesicle fusion in neurons (**Sweeney et al., 1995**). The cleavage site is not present in cellular brevin, the v-SNARE that facilitates vesicle release in other cell types. TNT is highly effective in small quantities—

any leakiness from the transgene can kill flies—and it may affect neurons below the GFP detection threshold; it lacks temporal control (Martin *et al.*, 2002). Also, there seem to be some cell types in which *UAS-TNT* is not effective (Rister and Heisenberg, 2006; Thum *et al.*, 2006); perhaps cellular brevin can compensate (Bhattacharya *et al.*, 2002). Recovery from TNT is also slow—the cell must synthesize new nSyb protein. But the bigger problem is that perdurance of small amounts of toxin may be sufficient to keep cleaving nSyb. There are several isotypes of botulinum, each of which produces a toxin that cleaves a different site in the SNARE complex; isotypes (B, D, F, and G) also cleave nSyb (Lalli *et al.*, 2003) which would make them neural specific as well, and perhaps these should be explored.

A revolution in the use of the UAS-GAL4 system for neural circuit mapping occurred when Kitamoto adapted a temperature-sensitive dominant negative mutation in *Shibire*, the *Drosophila* homolog of the vesicle recycling protein dynamin, to make a temporally and spatially controlled neural activity blocker **UAS-Shibire^{ts1}** (**Kitamoto, 2001**). Dynamin is a GTPase that forms rings around the necks of vesicles to pinch them off the membrane; the dominant negative version intercalates into the ring of proteins at the neck, but then it blocks the pinching activity (Danino and Hinshaw, 2001; Hinshaw, 2000). Electron microscopy of the *Shibire^{ts1}* mutation at the nonpermissive temperature shows a series of docked vesicles trapped on the membrane. The speed with which phenotypes manifest suggest that the first effects seen are neural, since neurons should be the cells most vulnerable to depletion of vesicles. Different neurons have different thresholds since they have different amounts of vesicle stores and different release demands. Large amounts of the dominant negative may be required in neurons that have large release zones or extensive vesicle reserves. *UAS-Shibire^{ts1}* has some effects—especially in muscles, for example— even at room temperature, and overexpression of wild-type dynamin can have effects. There is a constitutively dominant-negative *UAS-Shibire DN K44A* (Moline *et al.*, 1999) and a wild-type *UAS-Dynamin* (Entchev *et al.*, 2000) that are plausible controls. *UAS-Shibire^{ts1}* has been used successfully to identify neurons involved in courtship and memory retrieval as distinct from acquisition, a distinction that would not have been possible to make without the acute temporal control of neural activity that *UAS-Shibire^{ts1}* provides (**Broughton et al., 2004**; Dubnau *et al.*, 2001; Kitamoto, 2002; Waddell *et al.*, 2000; Wang *et al.*, 2003b).

UAS-TNT and *UAS-Shibire^{ts1}* act on small synaptic vesicle mediated chemical neurotransmission, which likely makes up the bulk of the information traffic in the fly brain, but they do not affect electrical transmission via gap junctions. If one sees an effect with these reagents, the neurons expressing them are implicated in the behavior, but the absence of an effect does not rule out a contribution from these neurons through electrical coupling. The "wireless network" of neuromodulators and peptides is critical for normal behavior. It is clear from work on the stomatogastric ganglia in crustaceans and on the leech locomotor choice (Briggman and Kristan, 2006; Friesen and Kristan, 2007; Marder and Bucher, 2007), for example, that the same circuit can produce very different outputs depending on the state of the system as set by neuromodulators. It remains to be proven how well reagents that work against fast neurotransmitter-containing synaptic vesicles block the release of dense-core vesicles containing **neuropeptides or neuromodulators** (Kaneko et al., 2000; McNabb et al., 1997).

C. Electrical blockers

Another strategy to inhibit neural function is to take advantage of a neuron's normal resting membrane potential and introduce a new "shunt" current that decreases excitability. There are several electrical shunts based on modified potassium channels. By increasing the permeability of the membrane at rest to potassium, one hyperpolarizes the resting membrane potential toward the equilibrium potential of potassium, thereby increasing the depolarization needed to fire action potentials. These potassium channel effectors will hyperpolarize muscles as well as neurons and are not readily reversible.

UAS-Kir2.1 (Baines et al., 2001; Paradis et al., 2001) is one such effector. It encodes a vertebrate inwardly rectifying potassium channel. Expression causes a hyperpolarized state and prevents neurons from depolarizing sufficiently to fire action potentials. It was tested in oocytes and at the neuromuscular junction, where its effects on membrane potential were directly measured. It has also been used to identify neurons involved in various behaviors in the adult, but the exact nature of the electrophysiological effect in these neurons has not been measured.

Expression of the **UAS-dOrk-deltaC** open rectifier potassium channel from *Drosophila* also reduces neural activity (Nitabach et al., 2002), again by hyperpolarizing the resting membrane potential, making it more difficult to depolarize and generate action potentials. This was confirmed in oocytes. There is a control construct that contains the dOrk channel with a pore mutation that destroys conductance (*UAS-dOrk-NC*) but does not seem to act as a dominant negative.

UAS-EKO, which stands for electrical knock out, is a truncated version of the Shaker voltage-sensitive potassium channel (White et al., 2001). It also reduces neural activity, but in this case by shortening action potentials and repolarizing neurons more rapidly. The mutations that were added to Shaker to make EKO shift the voltage activation threshold so that the channel opens with less membrane depolarization, so it is now closer to the time course of the sodium channel activation. Potassium channels normally cycle into an inactive state after opening; the ball and chain inactivation gate has been removed in EKO, allowing the channel to remain open. This speeds the repolarization of neurons and reduces their activity (Holmes et al., 2007). Expressing different levels of UAS-EKO can have a graded phenotypic effect.

A comparison of these potassium channel-based constructs in the circadian circuit found that UAS-Kir2.1 is stronger than UAS-dOrk, and both are much stronger than UAS-EKO (Holmes et al., 2007). It is not clear if this can be generalized to predict their relative strength in other neurons or muscles.

Tethered toxins from spiders and snails that act against neurotransmitter receptors and ion channels have been used in vertebrates (Ibanez-Tallon et al., 2004) and have been adapted for use in flies (Wu et al., 2008). The toxins are expressed in a given cell, secreted, and GPI anchored to act cell autonomously. Spider toxins that block inactivation of the primary sodium channel encoded by paralytic, the presynaptic calcium channel (cacophony), and a range of potassium channels have been tested for their effects on the PDF neurons that set the circadian clock. They have a variety of complicated effects on the membrane currents in these cells.

D. Neuronal activators

Circuit redundancy may make it difficult to identify what neurons are essential parts of a circuit simply by blocking their activity, especially with constitutively active reagents. Ectopically triggering behavior by activating neurons is an appealing complementary approach for identifying neurons sufficient to drive behavior. There are several options for increasing neural activity. **UAS-NaChBac** (Nitabach et al., 2006) based on a voltage gated bacterial sodium channel, increases activity in some neurons. NaChBac opens at −60 mV and should make cells more excitable. This may be an oversimplification: in some cells, UAS-NaChBac changes the cell from firing small but frequent action potentials to one producing bigger but less frequent spikes (Sheeba et al., 2008). NaChBac can also have side effects that look like inhibition: when expressed in bursicon-secreting neurons, it caused peptide depletion and the neurons were rendered inactive by the time bursicon release is required for cuticle tanning and wing expansion (Luan et al., 2006a).

Other neuronal activators are based on the Trp channels, which are nonselective cation channels that can be opened by ligands such as capsaicin, or by changes in temperature. The capsaicin-sensitive modified **VR1 Trp** channel (UAS-VR1E600K) has been used in worms (Tobin et al., 2002) and flies (Marella et al., 2006) to activate neurons in response to the capsaicin ligand. Overexpression of the Drosophila **TrpA1** channel renders neurons active upon temperature increase (Rosenzweig et al., 2008). Expression of the mammalian Trp channels may also prove effective. Some Trp channels can be activated at temperatures close to the physiological norm for flies and so may be acceptable for long-term behavioral studies.

In contrast to the potassium channel-based inhibitors described above, reducing potassium channel function should increase neural activity. Voltage-gated potassium channels are homotetramers. This makes it possible to transgenically express a faulty subunit that will intercalate into the tetramer and reduce or eliminate its channel function. Several of these dominant negative potassium channel constructs have been built by truncating the proteins between the N-terminal multimerization domain and the pore-forming transmembrane domains: **UAS-Eag-DN** (Broughton *et al.*, 2004), **UAS-Shaker-DN** (Mosca *et al.*, 2005), and **UAS-Shaw-DN** (Hodge *et al.*, 2005). These dominant negatives should constitutively block potassium channel function and should work only in neurons that normally express these channels.

E. Light-based methods

There are many strategies for using light to alter neural activity. Again, the primary development effort has been done in vertebrate systems. The fly-specific reagents are listed in Table 3.3. The phototransduction cascade itself is designed to convert photons into electrical activity and three proteins that make up this cascade in flies were expressed in vertebrate neurons to elicit neural depolarization in an approach called chARGe (Zemelman *et al.*, 2002). Light can be used to uncage a variety of neurotransmitters, second messengers, synthetic or natural ligands, or chemical modifiers. For example, the SPARK system uses an azobenzene chemical switch that can the bind to either an endogenous potassium channel or a modified channel that is transgenically expressed in specific populations of cells, making the channel light-gated (Banghart *et al.*, 2004; Fortin *et al.*, 2008; Zemelman *et al.*, 2003). Glutamate receptor activation can be triggered by light by expressing a modified LiGluR and using a light-activated glutamate agonist (Szobota *et al.*, 2007). When **UAS-P2X2**, encoding a vertebrate ATP-gated channel, is expressed in the fly giant fiber system and caged ATP is injected, the animals produce the characteristic jump-escape behavior in response to light (***Lima and Miesenbock, 2005***). Light can be used to alter cAMP levels with a photoactivatable adenylyl cyclase (Schroder-Lang *et al.*, 2007).

Another method attracting interest in *Drosophila* is Channelrhodopsin2 (**UAS-ChR2**), a monovalent cation channel from algae that is activated by ~470 nm blue light in the presence of all-trans retinal. ChR2 was cloned (Nagel *et al.*, 2003) and then adapted for neuroscience (Boyden *et al.*, 2005; Li *et al.*, 2005; Nagel *et al.*, 2005). The required retinal cofactor is present in mammalian neurons but not in *Drosophila*; it is generally supplied by feeding both during the larval and adult stages. Effective use of ChR2 may also require high transgene expression levels. There is now a red-shifted version of ChR2 from *Volvox* that is activated by light at 590 nm (Zhang *et al.*, 2008a). A new variant of ChR2 shows increased opening in response to a brief light pulse, potentially reducing the effect of blue light alone on neurons (Berndt et al., 2009). The two big problems here are light penetration and delivery of chemical cofactors. In some cases, the level of light required to activate the effectors also affects behavior or neural function directly. Ongoing development in this area is fast and hopefully reagents that are activated by longer wavelengths (which are less detectable by fly photoreceptors, penetrate farther into tissue, and might be amenable to two-photon excitation) and that do not require cofactors will be forthcoming.

Halorhodopsin **(NpHR)** is a chloride pump activated by ∼580 nm yellow light and also requires retinal as a cofactor. Halorhodopsin has also been engineered to express and reach the membrane at higher levels by adding an ER secretion signal (Gradinaru et al., 2008; Zhao et al., 2008). ChR2 and NpHR can be used together to generate specific firing patterns in neurons: they both have millisecond precision and are activated by different wavelengths of light (Zhang et al., 2007). There are frequent reviews on the state of "opto-genetics" (see Deisseroth et al., 2006; Miesenbock and Kevrekidis, 2005) and the intense interest in this highly promising field should continue to improve the reagents. For examples of this approach in flies, see Hwang et al. (2007), Schroll et al. (2006), and Suh et al. (2007).

F. Caveats

There are some cautionary notes for the use of all of the reagents that modify neural activity. First, the way these effectors act may be different in different cell types, and some reagents do not work in some cell types at all (Thum et al., 2006). In most cases, the electrophysiological characterization of these reagents was performed in an exogenous system, or at best in one particular cell type, and the validity of generalization is weak. The utility of the existing reagents on electrically coupled cells or peptidergic neurons is unclear. That makes positive results with the UAS-effectors more useful than negative results. (It is legitimate to say disrupting activity in these neurons alters behavior, but one should not conclude that expressing a given effector in a cell type and failing to see an effect really means those neurons are not involved.)

It is important to consider exactly what a given manipulation does to a neuron. For example, light on ChR2 flies triggers depolarization via an influx of cations. Blocking a potassium channel also increases neuronal excitability, but it does so by making it easier to fire action potentials triggered by the cells own sodium channels. It does not initiate firing de novo—it just amplifies an existing proclivity or lowers a threshold. These two "activators" might have very different effects on the same neural population.

The expression level of blockers can be critical—and the levels required may differ depending on the cell type. Each of the reporter and effector proteins has its own stability and effective concentration. We visualize the neurons in a

GAL4 pattern with one reporter, detect with antibodies, and then manipulate them with an effector such as *UAS-TNT*, which functions in very low doses, or one like *UAS-Shibire[ts1]*, which seems to require large amounts in many neurons. Do we see all of the neurons with GFP that express enough TNT to be impaired? Do we see many neurons with antibody-amplified GFP that do not make enough Shibire[ts1] to be blocked? Both false positives and false negatives are possible and worrying.

Timing of expression may also be critical: channel expression levels seem to be tightly controlled (Mee *et al.*, 2004) and homeostatic mechanisms work quickly to restore the ion balance (Turrigiano, 2008), so systems that rapidly trigger effector production or activity have the best chance of evading compensation. No one has systematically looked at the homeostatic responses of neurons expressing channel blockers or activators over time.

None of these cautions make the lines that alter neural activity invalid as tools for mapping which neurons are implicated in particular behaviors, but it does suggest that careful controls and comparison of the results with different blockers is prudent. Initially we can use the deadly effectors and complete blockers to winnow down which neurons are a key for particular behaviors. Then we can use reagents such as channelrhodopsin to send in signals that resemble those the neurons normally carry to more subtly alter the behavioral output and test our predictions about how neural circuits work to drive behavioral responses.

VI. QUANTITATIVE BEHAVIORAL ASSAYS

What can a fly do? What does a fly do? Developing ethologically relevant, quantifiable behavioral assays has been challenging and there are not that many established paradigms in the fly. For recent reviews of some behaviors studied in the fly, see Sharma *et al.* (2005) and Vosshall (2007). Table 3.4 includes a summary of most of the current behavioral paradigms studied in *Drosophila*. There is ample room for merging the behavior analysis techniques developed for other insects—and indeed other animals—with the genetic and molecular tools available in *Drosophila* to learn new things about how neural circuits drive behavior. For example, walking and searching assays from other insects could be adapted for flies (Buschges *et al.*, 2008; Merkle and Wehner, 2008; Watson *et al.*, 2002; Wittlinger *et al.*, 2006). Clever behavioral analysis alone can generate hypotheses about how neural circuits work. As an example, careful observation showed that emergency and voluntary flight initiation sequences are quite different, which suggests that different circuits underlie these behaviors. Further, the latency to jump is different when the emergency takeoff is initiated by a visual or an olfactory stimulus. Again, the circuits that mediate these responses may differ in more than just the sensory input layer

Table 3.4. Behaviors Commonly Assayed in *Drosophila*

Behavior	Description of behavior and usual assays	References
Locomotion		
General locomotion	It is possible to measure the number of lines crossed or to track flies in an open field. Larval crawling can also be tracked	Hughes and Thomas (2007) and Martin (2003)
Response to startle	Flies exhibit increased velocity in response to air puff, vibration, or odor delivery	Wolf *et al.* (2002)
Circadian rhythms	TriKinetics *Drosophila* activity monitoring system is an automatic beam cross detector that counts movement of individual flies in tubes; flies tend to be more active at dawn and dusk	Nitabach and Taghert (2008), Rosato and Kyriacou (2006), and Zordan *et al.* (2007)
Sleep	TriKinetics or ultrasound monitors watch for bouts of stillness lasting more than 5 min	Cirelli and Bushey (2008)
Righting reflex	Flies are knocked over and tested for their ability to get to their feet	Leal *et al.* (2004)
Jump-escape	Visual or olfactory stimuli can trigger an emergency takeoff, scored by hand or filmed with high-speed video	Trimarchi and Schneiderman (1995)
Flight	Flight can be measured by dropping flies into an oil-coated graduated cylinder where the best fliers stick initiate flight early and stick near the top or in flight arenas where tethered flies fly glued to a stick. Flight initiation can be studied with high-speed video	Benzer (1973), Card and Dickinson (2008a,b), Fry *et al.* (2008), Hammond and O'Shea (2007), Lehmann and Dickinson (2001), and Reiser and Dickinson (2008)
Landing response	Flies will extend their legs toward some objects rather than turning to avoid them in the flight arena	Maimon *et al.* (2008)
Gait analysis	A laser carpet measures foot placement and stride length	Strauss (1995)
Postural control	In an inebriometer, a series of baffles or funnels with slanted sides is used to test for lost of balance in response to ethanol or anesthetics. More resistant flies remain in the device longer, while susceptible ones elute more rapidly	Cohan and Hoffman (1986) and Moore *et al.* (1998)

(Continues)

Table 3.4. (*Continued*)

Behavior	Description of behavior and usual assays	References
Gap crossing	Flies will step across gaps of manageable size guided by visual cues.	Pick and Strauss (2005)
Leg resistance reflex	A leg can be moved manually and the required force measured	Ready et al. (1997)
Bang sensitivity	Flies usually recover rapidly from hard banging or vortex vibration but some mutants paralyze or seize	Pavlidis and Tanouye (1995)

Sensory

Behavior	Description of behavior and usual assays	References
Gravitaxis	Flies will move against gravity. This can be measured in a vertically oriented Y maze with multiple choice points	Baker et al. (2007a,b) and Kamicouchi et al. (2009)
Olfaction	An olfactory trap assay lures flies into funnels baited with different odorants; olfactory induced jump and T-maze choice tests can also be used to measure response to odorants. Olfactory and visual cues can be measured during flight. Larva crawls toward odor sources and track gradients	Frye and Dickinson (2004), Lilly and Carlson (1990), Louis et al. (2008b), McKenna et al. (1989), Suh et al. (2004), and Woodard et al. (1989)
Gustatory	Food consumption can be measured with liquid food in graduated capillary tubes; choice of egg laying sites can be used as a proxy for taste discrimination. Larva spends longer on certain food sources and their consumption can be measured by including dye in the food. They will avoid bitter food unless starved	Bader et al. (2007), Ja et al. (2007), Mery and Kawecki (2002), Wu et al. (2005), and Yang et al. (2008)
Phototaxis	The countercurrent device partitions flies based on how quickly they run toward light; more complicated visual motion tests can be performed in the flight arena with a virtual reality display on LED panels	Benzer (1967) and Reiser and Dickinson (2008)
Color vision or spectral preference	UV–visible light choice assays or motion stimuli in equiluminescent displays detect response to color	Gao et al. (2008) and Yamaguchi et al. (2008)
Optomotor response	Variants on the "fly stampede" measure response to optic flow or motion vision; flies will move against slow or sparse cues and with fast or dense stimuli	Katsov and Clandinin (2008)

(*Continues*)

Table 3.4. (*Continued*)

Behavior	Description of behavior and usual assays	References
Visual discrimination and persistence	Flies remember the location of an object that disappears from view and can recognize novel and familiar shapes in the flight area	Liu *et al.* (1999, 2006), Neuser *et al.* (2008), Peng *et al.* (2007), and Tang *et al.* (2004)
Audition	Playing courtship song to males induces male–male courtship; song induces females to slow down and accept copulation	Clyne and Miesenbock (2008), Crossley *et al.* (1995), Eberl *et al.* (1997), and Tauber and Eberl (2003)
Magnetosensation	A two choice maze with magnets shows flies sense magnetism	Gegear *et al.* (2008)
Thermosensation	Larva and adults avoid noxious heat and spend more time at preferred temperatures with <1 °C precision on a heated agar block	Hamada *et al.* (2008), Rosenzweig *et al.* (2005, 2008), Sayeed and Benzer (1996), Xu *et al.* (2006), and Zars (2001),
Mechanosensation	Larva back up in response to light touch; this can be measured by tapping them with an eyelash	Kernan *et al.* (1994)
Nociception or pain	Larva roll in response to heat or pinch	Tracey *et al.* (2003)
Hygrosensation	Flies can sense water	Hong *et al.* (2006), Inoshita and Tanimura (2006), and Liu *et al.* (2007)

Complex		
Male courtship	Insectavox measures courtship song production; other steps of courtship are usually scored manually from videotaped mating in small chambers	Gorcyzca and Hall (1987), Greenspan (2000), O'Dell (2003), Villella and Hall (2008), and Villella *et al.* (1997)
Female receptivity	Videotaped and scored manually; egg laying can serve as a proxy	Dickson (2008), Hasemeyer *et al.* (2009), Yang *et al.* (2009), and Yapici *et al.* (2008)
Grooming	Flies remove dust in a coordinated series of movements	Corfas and Dudai (1989, 1990) and Phillis *et al.* (1993)
Proboscis extension reflex	This is the motor program triggered by detection of an acceptable food source through tase bristles on the leg or labellum; it is observed and scored by hand	Shiraiwa and Carlson (2007)
Aggression	Male flies will perform a series of stereotyped movements when competing for resources such as food or mates. The behavior is filmed and scored manually	Chen *et al.* (2002), Dierick (2007), Hoyer *et al.* (2008), Mundiyanapurath *et al.* (2007), and Yurkovic *et al.* (2006)

(*Continues*)

Table 3.4. (*Continued*)

Behavior	Description of behavior and usual assays	References
Learning and memory		
Olfactory shock conditioning	Flies can be trained to avoid an odor associated with an electric shock in a T-maze	Berry *et al.* (2008) and Tully and Quinn (1985)
Habituation or sensitization	Flies respond differently to repeated presentation of sensory stimuli such as odor puffs or bristle touches	Asztalos *et al.* (2007), Engel and Wu (2008), and Joiner *et al.* (2007)
Spatial memory	Flies avoid the half of a box that has been associated with noxious heat	Diegelmann *et al.* (2006)
Courtship suppression	Males whose courtship attempts have been rejected are slower to court receptive females	Ejima *et al.* (2007) and Siegel and Hall (1979)
Appetitive conditioning	Larva and adults can also be conditioned to choose an odder associated with a reward	Fiala (2007) and Schroll *et al.* (2006)

(Card and Dickinson, 2008a,b; Hammond and O'Shea, 2007; Trimarchi and Schneiderman, 1995). The kinds of errors an animal makes also suggest the type or manner of computations it must be doing. It was deduced that motion vision in the fly uses local rather than global comparisons because of careful behavioral analysis of the errors the fly makes when the motion cues are distributed distantly across the ommatidia (Buchner, 1976). Careful behavioral analysis could be used to deepen our understanding of neural circuits in *Drosophila*.

Laboratory behavioral assays are designed to identify the neurons and genes that govern these behaviors in more natural environments. The goal is to understand the neural circuits that have evolved under natural selection and are optimized for what the flies actually do normally. We must be careful to consider how well our lab assays mimic the fly's normal circumstances and that we draw appropriate conclusions. The genetic background of the flies used in these assays and the neutrality of the markers we use to identify transgenes must also be considered. For example, there is literature to suggest that using *mini-white* to mark transgenes that are then assayed for courtship defects can lead to results that are difficult to interpret (An *et al.*, 2000; Borycz *et al.*, 2008; Zhang and Odenwald, 1995). Some behaviors, including locomotion speeds, aggression, and response to magnetic fields, also seem to depend on the genetic background of the flies (Gegear *et al.*, 2008; Hoyer *et al.*, 2008; Wolf *et al.*, 2002). The advantage of using isogenic lines or F1 hybrids is also under debate (Sharma *et al.*, 2005). These caveats are manageable as long as appropriate controls are performed.

VII. CONCLUSIONS

A. Example circuits

A few examples are discussed below to highlight the excellent use of the genetic tools available for circuit analysis in *Drosophila* and to describe the circuits about which we know the most. Beautiful work on olfaction has shown organizational logic of the first order sensory representation maps in the antennal lobe (Couto *et al.*, 2005; Laissue *et al.*, 1999; Marella *et al.*, 2006; Marin *et al.*, 2002; Vosshall *et al.*, 2000; Wang *et al.*, 2002, 2003a,b). Excitatory and inhibitory interneurons modify the representation of odors within the antennal lobe (Olsen and Wilson, 2008b; Olsen *et al.*, 2007; Shang *et al.*, 2007), and projection neurons carry this information on to the lateral horn and the Kenyon cells of the mushroom bodies (Jefferis *et al.*, 2007; Murthy *et al.*, 2008; Turner *et al.*, 2008). Olfactory information is integrated with visual cues (Frye *et al.*, 2003), but where and how this occurs is unknown. In the larva, how the sensory input affects the behavioral response has also been studied (Kreher *et al.*, 2008; Louis *et al.*, 2008a). How the odor sensory experience is connected to the motor neurons that dictate behavioral response is also unknown.

Work on the gustatory system has shown that the sensory neurons carrying different taste qualities project to different areas of the subesophogeal ganglia (Marella *et al.*, 2006; Fischler *et al.*, 2007; Wang *et al.*, 2004b). Some of the motor neurons that drive the extension of the proboscis in response to an attractive taste have been identified, but the sensory neurons do not directly connect to the motor neurons, revealing that this apparently simple reflex has additional circuitry that remains to be uncovered (Gordon and Scott, 2009a,b). Part of a circuit governing response to taste has also been established in the larva (Bader *et al.*, 2007; Melcher and Pankratz, 2005).

Extensive work in the visual system has identified neurons involved in detecting motion and color; these are reviewed in Borst (2009). For recent advances here, see Gao *et al.* (2008), Joesch *et al.* (2008), Katsov and Clandinin (2008), Morante and Desplan (2008), Rister *et al.* (2007), and Yamaguchi *et al.* (2008). Very little is known about auditory processing beyond the sensory apparatus in the Johnston's organ (Kernan, 2007). A careful anatomical analysis of projections from the Johnston's organ has been conducted and the function and destination of these projections is under investigation (Kamikouchi *et al.*, 2006, 2009; Yorozu *et al.*, 2009). The sensory systems have been intensively studied and have seen great progress, but in no case can we trace a circuit all the way from a sensory stimulus to a motor output.

There are many examples where parts of a circuit have been mapped. The giant fiber circuit that mediates the jump-escape response has been studied by mutagenesis screens, anatomical dye fills, and electrophysiology. The giant fibers and the motor neurons that drive the jump are known, but the neurons that activate the giant fibers in the brain are not (Allen et al., 2006). Some of the neurons that govern circadian rhythms and light mediated arousal have been identified (Nitabach and Taghert, 2008; Shang et al., 2008; Sheeba et al., 2008). The neuropeptide-releasing neurons that initiate the wing expansion after eclosion have been determined using an arsenal of intersectional genetic strategies (Luan et al., 2006a). Neurons in and near the mushroom bodies that govern olfactory memory storage and retrieval have also been characterized, reviewed in Keene and Waddell (2007). Perhaps the most complex circuit under intensive study is the network of neurons that underlie male courtship behavior (Clyne and Miesenbock, 2008; Datta et al., 2008; Kimura et al., 2005, 2008; Manoli et al., 2005; Stockinger et al., 2005). The male-specific isoform of *fruitless* is expressed in sensory, motor, and interneurons that participate in courtship. One of the interesting, unanswered questions here is how much of the difference between male and female courtship behavior is reflected in anatomical differences in the circuitry and how much is due to gender differences in the activity or synaptic weights within anatomically similar circuitry. *Fruitless* might be a "master control gene for behavior" (Yamamoto, 2007) in much the same way that there are hypothesized to be "command neurons" (Weiss and Kupfermann, 1978) that trigger entire behavioral programs. *Fruitless* may be unusual in that it is expressed in many types of neurons that participate in courtship, but even here, there are still missing circuit elements.

B. New tools

To highlight some of the newest developments in the arsenal of tools for circuit mapping in flies, the tethered toxins that block specific ion channels and the Trp channels that activate neurons in response to temperature changes seem very promising. The new generation of GECIs for monitoring neural activity and the positive intersectional methods for narrowing gene expression should also contribute to our ability to refine the maps of which neurons participate in given behaviors.

We can learn from the techniques being developed in vertebrate and nematode systems. A variety of strategies for neuronal activation and inactivation exist there: MIST (Karpova et al., 2005), Allatostatin (Tan et al., 2006), Ivermectin (Lerchner et al., 2007), RASSLs and DREADDs (G-protein coupled receptors with synthetic ligands: Armbruster et al. (2007) and Conklin et al. (2008)), and modified GABA receptors (Wulff et al., 2007) have all been used for circuit mapping. The use of Cre recombinase lines for positive intersectional approaches to transgene expression and the idea that reporters can be recombined out after stable lines have been established are common practice in transgenic mouse work (Dymecki and Kim, 2007). Channelrhodopsin has been used for anatomical mapping as well as for linking neural activity to behavioral output (Arenkiel *et al.*, 2007; Petreanu *et al.*, 2007; Wang *et al.*, 2007). This mapping was done with electrodes but there is the possibility that it could be done with optical reporters of neural activity (Airan *et al.*, 2007). For visualizing large numbers of neurons at once and tracing their connections, the Brainbow technique (Lichtman *et al.*, 2008; Livet *et al.*, 2007) and the self-amplifying viral trans-synaptic tracers invite envy (Wickersham *et al.*, 2007). Watching neural circuits in action in a behaving animal has been accomplished in several genetic model organisms (Clark *et al.*, 2007; Dombeck *et al.*, 2007; Faumont and Lockery, 2006; Orger *et al.*, 2008; Zhang *et al.*, 2008b). In *C. elegans*, the GRASP synapse marking strategy and the complete electron microscopy wiring diagram suggest what we could do with their equivalents in the fly (Chalasani *et al.*, 2007; Chen *et al.*, 2006; Feinberg *et al.*, 2008; Gray *et al.*, 2005).

C. Full circle

When the field of "neurogenetics" was born, there was debate about whether it would be possible to find single gene mutations that led to understanding behaviors (Vosshall, 2007). Now it is widely accepted that the genetic approach has yielded insights into behavior, especially in the identification of genes involved in neurodevelopment, axon wiring, and neural function. Some of the early stars of neurogenetics were the conditional mutations that led to the cloning of ion channels and synaptic vesicle release machinery; these same mutations allowed the design of tools to manipulate neural activity. To understand neural circuits, we are now using genetic tools to dissect which neurons play roles in specific behaviors, how these neurons are linked, and what jobs they do.

The correlative tools to observe neural activity—electrophysiologically or optically—while the animal is receiving some sensory stimulation or performing some motor behavior can now be combined with the causative genetic manipulations which change activity in defined populations in neurons to see what behavioral changes they evoke. The powerful genetic toolkit that makes *Drosophila* famous can now be applied creatively to address how neural circuits are organized to control appropriate behavioral responses to the environment and experience of the animal. The general principles determined by understanding circuit logic in the specific cases in the fruit fly should be informative for other systems as well.

Acknowledgments

I thank Gerry Rubin, Steven Goodwin, Loren Looger, Julide Bilen, Claire McKellar, Andrew Seeds, and Arnim Jenett for helpful comments on the manuscript. Special thanks go to Vivek Jayaraman and Stephanie Albin for help with the electrophysiology section, Michael Reiser for advice on the behavior table, and the Howard Hughes Medical Institute for funding.

References

Airan, R. D., Hu, E. S., Vijaykumar, R., Roy, M., Meltzer, L. A., and Deisseroth, K. (2007). Integration of light-controlled neuronal firing and fast circuit imaging. *Curr. Opin. Neurobiol.* **17**, 587–592.

Alawi, A. A., and Pak, W. L. (1971). On-transient of insect electroretinogram: Its cellular origin. *Science (New York, N.Y.)* **172**, 1055–1057.

Allen, M. J., Shan, X., Caruccio, P., Froggett, S. J., Moffat, K. G., and Murphey, R. K. (1999). Targeted expression of truncated glued disrupts giant fiber synapse formation in *Drosophila*. *J. Neurosci.* **19**, 9374–9384.

Allen, M. J., O'Kane, C. J., and Moffat, K. G. (2002). Cell ablation using wild-type and cold-sensitive ricin-A chain in *Drosophila* embryonic mesoderm. *Genesis* **34**, 132–134.

Allen, M. J., Godenschwege, T. A., Tanouye, M. A., and Phelan, P. (2006). Making an escape: Development and function of the *Drosophila* giant fibre system. *Semin. Cell Dev. Biol.* **17**, 31–41.

An, X., Armstrong, J. D., Kaiser, K., and O'Dell, K. M. (2000). The effects of ectopic white and transformer expression on *Drosophila* courtship behavior. *J. Neurogenet.* **14**, 227–243, 271.

Andersen, R., Li, Y., Resseguie, M., and Brenman, J. E. (2005). Calcium/calmodulin-dependent protein kinase II alters structural plasticity and cytoskeletal dynamics in *Drosophila*. *J. Neurosci.* **25**, 8878–8888.

Arenkiel, B. R., Peca, J., Davison, I. G., Feliciano, C., Deisseroth, K., Augustine, G. J., Ehlers, M. D., and Feng, G. (2007). *In vivo* light-induced activation of neural circuitry in transgenic mice expressing channelrhodopsin-2. *Neuron* **54**, 205–218.

Armbruster, B. N., Li, X., Pausch, M. H., Herlitze, S., and Roth, B. L. (2007). Evolving the lock to fit the key to create a family of G protein-coupled receptors potently activated by an inert ligand. *Proc. Natl. Acad. Sci. USA* **104**, 5163–5168.

Armstrong, J. D., Texada, M. J., Munjaal, R., Baker, D. A., and Beckingham, K. M. (2006). Gravitaxis in *Drosophila* melanogaster: A forward genetic screen. *Genes Brain Behav.* **5**, 222–239.

Asahina, K., Louis, M., Piccinotti, S., and Vosshall, L. B. (2009). A circuit supporting concentration-invariant odor perception in *Drosophila*. *J. Biol.* **8**, 9.

Asztalos, Z., Arora, N., and Tully, T. (2007). Olfactory jump reflex habituation in *Drosophila* and effects of classical conditioning mutations. *J. Neurogenet.* **21**, 1–18.

Ataka, K., and Pieribone, V. A. (2002). A genetically targetable fluorescent probe of channel gating with rapid kinetics. *Biophys. J.* **82**, 509–516.

Bader, R., Colomb, J., Pankratz, B., Schrock, A., Stocker, R. F., and Pankratz, M. J. (2007). Genetic dissection of neural circuit anatomy underlying feeding behavior in *Drosophila*: Distinct classes of hugin-expressing neurons. *J. Comp. Neurol.* **502**, 848–856.

Baines, R. A., and Bate, M. (1998). Electrophysiological development of central neurons in the *Drosophila* embryo. *J. Neurosci.* **18**, 4673–4683.

Baines, R. A., and Pym, E. C. (2006). Determinants of electrical properties in developing neurons. *Semin. Cell Dev. Biol.* **17**, 12–19.

Baines, R. A., Uhler, J. P., Thompson, A., Sweeney, S. T., and Bate, M. (2001). Altered electrical properties in *Drosophila* neurons developing without synaptic transmission. *J. Neurosci.* **21**, 1523–1531.

Baird, G. S., Zacharias, D. A., and Tsien, R. Y. (1999). Circular permutation and receptor insertion within green fluorescent proteins. *Proc. Natl. Acad. Sci. USA* **96**, 11241–11246.

Baker, B. S., Taylor, B. J., and Hall, J. C. (2001). Are complex behaviors specified by dedicated regulatory genes? Reasoning from Drosophila. *Cell* **105**, 13–24.

Baker, B. J., Lee, H., Pieribone, V. A., Cohen, L. B., Isacoff, E. Y., Knopfel, T., and Kosmidis, E. K. (2007a). Three fluorescent protein voltage sensors exhibit low plasma membrane expression in mammalian cells. *J. Neurosci. Methods* **161**, 32–38.

Baker, D. A., Beckingham, K. M., and Armstrong, J. D. (2007b). Functional dissection of the neural substrates for gravitaxic maze behavior in *Drosophila* melanogaster. *J. Comp. Neurol.* **501,** 756–764.

Baker, B. J., Mutoh, H., Dimitrov, D., Akemann, W., Perron, A., Iwamoto, Y., Jin, L., Cohen, L. B., Isacoff, E. Y., Pieribone, V. A., *et al.* (2008). Genetically encoded fluorescent sensors of membrane potential. *Brain Cell Biol.* **36,** 53–67.

Banghart, M., Borges, K., Isacoff, E., Trauner, D., and Kramer, R. H. (2004). Light-activated ion channels for remote control of neuronal firing. *Nat. Neurosci.* **7,** 1381–1386.

Barnea, G., Strapps, W., Herrada, G., Berman, Y., Ong, J., Kloss, B., Axel, R., and Lee, K. J. (2008). The genetic design of signaling cascades to record receptor activation. *Proc. Natl. Acad. Sci. USA* **105,** 64–69.

Barth, A. L., Gerkin, R. C., and Dean, K. L. (2004). Alteration of neuronal firing properties after in vivo experience in a FosGFP transgenic mouse. *J. Neurosci.* **24,** 6466–6475.

Basler, K., and Struhl, G. (1994). Compartment boundaries and the control of *Drosophila* limb pattern by hedgehog protein. *Nature* **368,** 208–214.

Belvin, M. P., Zhou, H., and Yin, J. C. (1999). The *Drosophila* dCREB2 gene affects the circadian clock. *Neuron* **22,** 777–787.

Benzer, S. (1967). Behavioral mutants of *Drosophila* isolated by countercurrent distribution. *Proc. Natl. Acad. Sci. USA* **58,** 1112–1119.

Benzer, S. (1973). Genetic dissection of behavior. *Sci. Am.* **229,** 24–37.

Berndt, A., Yizhar, O., Gunaydin, L. A., Hegemann, P., and Deisseroth, K. (2009). Bi-stable neural state switches. *Nat. Neurosci.* **12,** 229–234.

Berry, J., Krause, W. C., and Davis, R. L. (2008). Olfactory memory traces in *Drosophila*. *Prog. Brain Res.* **169,** 293–304.

Bhandawat, V., Olsen, S. R., Gouwens, N. W., Schlief, M. L., and Wilson, R. I. (2007). Sensory processing in the *Drosophila* antennal lobe increases reliability and separability of ensemble odor representations. *Nat. Neurosci.* **10,** 1474–1482.

Bhattacharya, S., Stewart, B. A., Niemeyer, B. A., Burgess, R. W., McCabe, B. D., Lin, P., Boulianne, G., O'Kane, C. J., and Schwarz, T. L. (2002). Members of the synaptobrevin/vesicle-associated membrane protein (VAMP) family in *Drosophila* are functionally interchangeable in vivo for neurotransmitter release and cell viability. *Proc. Natl. Acad. Sci. USA* **99,** 13867–13872.

Bicker, G. (1999). Introduction to neurotransmitter histochemistry of the insect brain. *Microsc. Res. Tech.* **45,** 63–64.

Bischof, J., and Basler, K. (2008). Recombinases and their use in gene activation, gene inactivation, and transgenesis. *Methods Mol. Biol. (Clifton, N.J.)* **420,** 175–195.

Bischof, J., Maeda, R. K., Hediger, M., Karch, F., and Basler, K. (2007). An optimized transgenesis system for *Drosophila* using germ-line-specific phiC31 integrases. *Proc. Natl. Acad. Sci. USA* **104,** 3312–3317.

Blair, S. S. (2003). Genetic mosaic techniques for studying *Drosophila* development. *Development (Cambridge, England)* **130,** 5065–5072.

Borst, A. (1984). Identification of different chemoreceptors by electroantennogram recording. *J. Insect Physiol.* 507–510.

Borst, A. (2009). *Drosophila*'s view on insect vision. *Curr. Biol.* **19,** R36–R47.

Borycz, J., Borycz, J. A., Kubow, A., Lloyd, V., and Meinertzhagen, I. A. (2008). *Drosophila* ABC transporter mutants white, brown and scarlet have altered contents and distribution of biogenic amines in the brain. *J. Exp. Biol.* **211,** 3454–3466.

Boyden, E. S., Zhang, F., Bamberg, E., Nagel, G., and Deisseroth, K. (2005). Millisecond-timescale, genetically targeted optical control of neural activity. *Nat. Neurosci.* **8,** 1263–1268.

Brand, A. H., and Perrimon, N. (1993). Targeted gene expression as a means of altering cell fates and generating dominant phenotypes. *Development (Cambridge, England)* **118,** 401–415.

Briggman, K. L., and Kristan, W. B., Jr. (2006). Imaging dedicated and multifunctional neural circuits generating distinct behaviors. *J. Neurosci.* **26**, 10925–10933.

Briggman, K. L., Abarbanel, H. D., and Kristan, W. B., Jr. (2005). Optical imaging of neuronal populations during decision-making. *Science (New York, N.Y.)* **307**, 896–901.

Broadie, K. (2000a). Electrophysiological approaches to the neuromusculature. In *Drosophila* Protocols, (W. Sullivan, M. Ashburner, and R. S. Hawley, eds.), pp. 273–296. Cold Spring Harbor Laboratory Press, Cold Spring Harbor, NY.

Broadie, K. (2000b). Functional assays of the peripheral and central nervous systems. In *Drosophila* Protocols, (W. Sullivan, M. Ashburner, and R. S. Hawley, eds.), pp. 297–312. Cold Spring Harbor Laboratory Press, Cold Spring Harbor, NY.

Broadie, K., and Bate, M. (1993). Activity-dependent development of the neuromuscular synapse during *Drosophila* embryogenesis. *Neuron* **11**, 607–619.

Broughton, S. J., Kitamoto, T., and Greenspan, R. J. (2004). Excitatory and inhibitory switches for courtship in the brain of *Drosophila* melanogaster. *Curr. Biol.* **14**, 538–547.

Buchner, E. (1976). Elementary visual detectors in and insect visual system. *Biol. Cybern.* **24**, 85–101.

Bulina, M. E., Chudakov, D. M., Britanova, O. V., Yanushevich, Y. G., Staroverov, D. B., Chepurnykh, T. V., Merzlyak, E. M., Shkrob, M. A., Lukyanov, S., and Lukyanov, K. A. (2006). A genetically encoded photosensitizer. *Nat. Biotechnol.* **24**, 95–99.

Burrone, J. (2005). Synaptic physiology: Illuminating the road ahead. *Curr. Biol.* **15**, R876–R878.

Buschges, A., Akay, T., Gabriel, J. P., and Schmidt, J. (2008). Organizing network action for locomotion: Insights from studying insect walking. *Brain Res. Rev.* **57**, 162–171.

Cachero, S., and Jefferis, G. S. (2008). *Drosophila* olfaction: The end of stereotypy? *Neuron* **59**, 843–845.

Callahan, C. A., and Thomas, J. B. (1994). Tau-beta-galactosidase, an axon-targeted fusion protein. *Proc. Natl. Acad. Sci. USA* **91**, 5972–5976.

Cao, G., and Nitabach, M. N. (2008). Circadian control of membrane excitability in *Drosophila* melanogaster lateral ventral clock neurons. *J. Neurosci.* **28**, 6493–6501.

Card, G., and Dickinson, M. (2008a). Performance trade-offs in the flight initiation of *Drosophila*. *J. Exp. Biol.* **211**, 341–353.

Card, G., and Dickinson, M. H. (2008b). Visually mediated motor planning in the escape response of *Drosophila*. *Curr. Biol.* **18**, 1300–1307.

Chalasani, S. H., Chronis, N., Tsunozaki, M., Gray, J. M., Ramot, D., Goodman, M. B., and Bargmann, C. I. (2007). Dissecting a circuit for olfactory behaviour in Caenorhabditis elegans. *Nature* **450**, 63–70.

Chanda, B., Blunck, R., Faria, L. C., Schweizer, F. E., Mody, I., and Bezanilla, F. (2005). A hybrid approach to measuring electrical activity in genetically specified neurons. *Nat. Neurosci.* **8**, 1619–1626.

Chen, S., Lee, A. Y., Bowens, N. M., Huber, R., and Kravitz, E. A. (2002). Fighting fruit flies: A model system for the study of aggression. *Proc. Natl. Acad. Sci. USA* **99**, 5664–5668.

Chen, B. L., Hall, D. H., and Chklovskii, D. B. (2006). Wiring optimization can relate neuronal structure and function. *Proc. Natl. Acad. Sci. USA* **103**, 4723–4728.

Chiang, A. S., Blum, A., Barditch, J., Chen, Y. H., Chiu, S. L., Regulski, M., Armstrong, J. D., Tully, T., and Dubnau, J. (2004). Radish encodes a phospholipase-A2 and defines a neural circuit involved in anesthesia-resistant memory. *Curr. Biol.* **14**, 263–272.

Choi, J. C., Park, D., and Griffith, L. C. (2004). Electrophysiological and morphological characterization of identified motor neurons in the *Drosophila* third instar larva central nervous system. *J. Neurophysiol.* **91**, 2353–2365.

Cirelli, C., and Bushey, D. (2008). Sleep and wakefulness in *Drosophila* melanogaster. *Ann. N. Y. Acad. Sci.* **1129**, 323–329.

Clark, I. E., Jan, L. Y., and Jan, Y. N. (1997). Reciprocal localization of Nod and kinesin fusion proteins indicates microtubule polarity in the *Drosophila* oocyte, epithelium, neuron and muscle. *Development (Cambridge, England)* **124,** 461–470.

Clark, D. A., Gabel, C. V., Gabel, H., and Samuel, A. D. (2007). Temporal activity patterns in thermosensory neurons of freely moving Caenorhabditis elegans encode spatial thermal gradients. *J. Neurosci.* **27,** 6083–6090.

Clyne, J. D., and Miesenbock, G. (2008). Sex-specific control and tuning of the pattern generator for courtship song in *Drosophila*. *Cell* **133,** 354–363.

Clyne, P., Grant, A., O'Connell, R., and Carlson, J. R. (1997). Odorant response of individual sensilla on the *Drosophila* antenna. *Invert. Neurosci.* **3,** 127–135.

Cohan, F. M., and Hoffmann, A. A. (1986). Genetic divergence under uniform selection. II. Different responses to selection for knockdown resistance to ethanol among *Drosophila* melanogaster populations and their replicate lines. *Genetics* **114,** 145–164.

Conklin, B. R., Hsiao, E. C., Claeysen, S., Dumuis, A., Srinivasan, S., Forsayeth, J. R., Guettier, J. M., Chang, W. C., Pei, Y., McCarthy, K. D., *et al.* (2008). Engineering GPCR signaling pathways with RASSLs. *Nat. Methods* **5,** 673–678.

Connolly, J. B., Roberts, I. J., Armstrong, J. D., Kaiser, K., Forte, M., Tully, T., and O'Kane, C. J. (1996). Associative learning disrupted by impaired Gs signaling in *Drosophila* mushroom bodies. *Science (New York, N.Y.)* **274,** 2104–2107.

Corfas, G., and Dudai, Y. (1989). Habituation and dishabituation of a cleaning reflex in normal and mutant *Drosophila*. *J. Neurosci.* **9,** 56–62.

Corfas, G., and Dudai, Y. (1990). Adaptation and fatigue of a mechanosensory neuron in wild-type *Drosophila* and in memory mutants. *J. Neurosci.* **10,** 491–499.

Couto, A., Alenius, M., and Dickson, B. J. (2005). Molecular, anatomical, and functional organization of the *Drosophila* olfactory system. *Curr. Biol.* **15,** 1535–1547.

Crossley, S. A., Bennet-Clark, H. C., and Evert, H. T. (1995). Courtship song components affect male and female *Drosophila* differently. *Anim. Behav.* **50,** 827–839.

Damasio, H., Grabowski, T., Frank, R., Galaburda, A. M., and Damasio, A. R. (1994). The return of Phineas Gage: Clues about the brain from the skull of a famous patient. *Science (New York, N.Y.)* **264,** 1102–1105.

Danino, D., and Hinshaw, J. E. (2001). Dynamin family of mechanoenzymes. *Curr. Opin. Cell Biol.* **13,** 454–460.

Dankert, H., Wang, L., Hoopfer, E. D., Anderson, D. J., and Perona, P. (2009). Automated monitoring and analysis of social behavior in Drosophila. *Nat. Methods* **6,** 297–303.

Datta, S. R., Vasconcelos, M. L., Ruta, V., Luo, S., Wong, A., Demir, E., Flores, J., Balonze, K., Dickson, B. J., and Axel, R. (2008). The *Drosophila* pheromone cVA activates a sexually dimorphic neural circuit. *Nature* **452,** 473–477.

de Belle, J. S., and Heisenberg, M. (1994). Associative odor learning in *Drosophila* abolished by chemical ablation of mushroom bodies. *Science (New York, N.Y.)* **263,** 692–695.

de Belle, J. S., Hilliker, A. J., and Sokolowski, M. B. (1989). Genetic localization of foraging (for): A major gene for larval behavior in *Drosophila* melanogaster. *Genetics* **123,** 157–163.

de Bruyne, M., Clyne, P. J., and Carlson, J. R. (1999). Odor coding in a model olfactory organ: The *Drosophila* maxillary palp. *J. Neurosci.* **19,** 4520–4532.

Deisseroth, K., Feng, G., Majewska, A. K., Miesenbock, G., Ting, A., and Schnitzer, M. J. (2006). Next-generation optical technologies for illuminating genetically targeted brain circuits. *J. Neurosci.* **26,** 10380–10386.

Demir, E., and Dickson, B. J. (2005). fruitless splicing specifies male courtship behavior in *Drosophila*. *Cell* **121,** 785–794.

Dickinson, M. H., and Palka, J. (1987). Physiological properties, time of development, and central projection are correlated in the wing mechanoreceptors of *Drosophila*. *J. Neurosci.* **7,** 4201–4208.

Dickson, B. J. (2008). Wired for sex: The neurobiology of Drosophila mating decisions. Science (New York, N.Y.) 322, 904–909.

Diegelmann, S., Zars, M., and Zars, T. (2006). Genetic dissociation of acquisition and memory strength in the heat-box spatial learning paradigm in Drosophila. Learn. Mem. (Cold Spring Harbor, N.Y.) 13, 72–83.

Dierick, H. A. (2007). A method for quantifying aggression in male Drosophila melanogaster. Nat. Protoc. 2, 2712–2718.

Dietzl, G., Chen, D., Schnorrer, F., Su, K. C., Barinova, Y., Fellner, M., Gasser, B., Kinsey, K., Oppel, S., Scheiblauer, S., et al. (2007). A genome-wide transgenic RNAi library for conditional gene inactivation in Drosophila. Nature 448, 151–156.

Dombeck, D. A., Khabbaz, A. N., Collman, F., Adelman, T. L., and Tank, D. W. (2007). Imaging large-scale neural activity with cellular resolution in awake, mobile mice. Neuron 56, 43–57.

Dubnau, J., Grady, L., Kitamoto, T., and Tully, T. (2001). Disruption of neurotransmission in Drosophila mushroom body blocks retrieval but not acquisition of memory. Nature 411, 476–480.

Duffy, J. B. (2002). GAL4 system in Drosophila: A fly geneticist's Swiss army knife. Genesis 34, 1–15.

Dymecki, S. M., and Kim, J. C. (2007). Molecular neuroanatomy's "Three Gs": A primer. Neuron 54, 17–34.

Eberl, D. F., Duyk, G. M., and Perrimon, N. (1997). A genetic screen for mutations that disrupt an auditory response in Drosophila melanogaster. Proc. Natl. Acad. Sci. USA 94, 14837–14842.

Ejima, A., Smith, B. P., Lucas, C., van der Goes van Naters, W., Miller, C. J., Carlson, J. R., Levine, J. D., and Griffith, L. C. (2007). Generalization of courtship learning in Drosophila is mediated by cis-vaccenyl acetate. Curr. Biol. 17, 599–605.

Elkins, T., and Ganetzky, B. (1990). Conduction in the giant nerve fiber pathway in temperature-sensitive paralytic mutants of Drosophila. J. Neurogenet. 6, 207–219.

Engel, J. E., and Wu, C. F. (1996). Altered habituation of an identified escape circuit in Drosophila memory mutants. J. Neurosci. 16, 3486–3499.

Engel, J. E., and Wu, C. F. (2008). Neurogenetic approaches to habituation and dishabituation in Drosophila. Neurobiol. Learn Mem.

Entchev, E. V., Schwabedissen, A., and Gonzalez-Gaitan, M. (2000). Gradient formation of the TGF-beta homolog Dpp. Cell 103, 981–991.

Eresh, S., Riese, J., Jackson, D. B., Bohmann, D., and Bienz, M. (1997). A CREB-binding site as a target for decapentaplegic signalling during Drosophila endoderm induction. EMBO J. 16, 2014–2022.

Estes, P. S., Ho, G. L., Narayanan, R., and Ramaswami, M. (2000). Synaptic localization and restricted diffusion of a Drosophila neuronal synaptobrevin–green fluorescent protein chimera in vivo. J. Neurogenet. 13, 233–255.

Faumont, S., and Lockery, S. R. (2006). The awake behaving worm: Simultaneous imaging of neuronal activity and behavior in intact animals at millimeter scale. J. Neurophysiol. 95, 1976–1981.

Fayyazuddin, A., Zaheer, M. A., Hiesinger, P. R., and Bellen, H. J. (2006). The nicotinic acetylcholine receptor Dalpha7 is required for an escape behavior in Drosophila. PLoS Biol. 4, e63.

Feinberg, E. H., Vanhoven, M. K., Bendesky, A., Wang, G., Fetter, R. D., Shen, K., and Bargmann, C. I. (2008). GFP reconstitution across synaptic partners (GRASP) defines cell contacts and synapses in living nervous systems. Neuron 57, 353–363.

Ferris, J., Ge, H., Liu, L., and Roman, G. (2006). G(o) signaling is required for Drosophila associative learning. Nat. Neurosci. 9, 1036–1040.

Fiala, A. (2007). Olfaction and olfactory learning in Drosophila: Recent progress. Curr. Opin. Neurobiol. 17, 720–726.

Fiala, A., and Spall, T. (2003). In vivo calcium imaging of brain activity in Drosophila by transgenic cameleon expression. Sci STKE PL62003.

Fiala, A., Spall, T., Diegelmann, S., Eisermann, B., Sachse, S., Devaud, J. M., Buchner, E., and Galizia, C. G. (2002). Genetically expressed cameleon in Drosophila melanogaster is used to visualize olfactory information in projection neurons. Curr. Biol. 12, 1877–1884.

Fischbach, K. F., and Dittrich, A. P. (1989). The optic lobe of Drosophila melanogaster. I. A Golgi analysis of wild-type structure. Cell Tissue Res. 258, 441–475.

Fischer, J. A., Giniger, E., Maniatis, T., and Ptashne, M. (1988). GAL4 activates transcription in Drosophila. Nature 332, 853–856.

Fischler, W., Kong, P., Marella, S., and Scott, K. (2007). The detection of carbonation by the Drosophila gustatory system. Nature 448, 1054–1057.

Fish, M. P., Groth, A. C., Calos, M. P., and Nusse, R. (2007). Creating transgenic Drosophila by microinjecting the site-specific phiC31 integrase mRNA and a transgene-containing donor plasmid. Nat. Protoc. 2, 2325–2331.

Fortin, D. L., Banghart, M. R., Dunn, T. W., Borges, K., Wagenaar, D. A., Gaudry, Q., Karakossian, M. H., Otis, T. S., Kristan, W. B., Trauner, D., and Kramer, R. H. (2008). Photochemical control of endogenous ion channels and cellular excitability. Nat. Methods 5, 331–338.

Friesen, W. O., and Kristan, W. B. (2007). Leech locomotion: Swimming, crawling, and decisions. Curr. Opin. Neurobiol. 17, 704–711.

Friggi-Grelin, F., Coulom, H., Meller, M., Gomez, D., Hirsh, J., and Birman, S. (2003). Targeted gene expression in Drosophila dopaminergic cells using regulatory sequences from tyrosine hydroxylase. J. Neurobiol. 54, 618–627.

Fry, S. N., Rohrseitz, N., Straw, A. D., and Dickinson, M. H. (2008). TrackFly: Virtual reality for a behavioral system analysis in free-flying fruit flies. J. Neurosci. Methods 171, 110–117.

Frye, M. A., and Dickinson, M. H. (2004). Motor output reflects the linear superposition of visual and olfactory inputs in Drosophila. J. Exp. Biol. 207, 123–131.

Frye, M. A., Tarsitano, M., and Dickinson, M. H. (2003). Odor localization requires visual feedback during free flight in Drosophila melanogaster. J. Exp. Biol. 206, 843–855.

Gallo, S. M., Li, L., Hu, Z., and Halfon, M. S. (2006). REDfly: A regulatory element database for Drosophila. Bioinformatics 22, 381–383.

Gao, S., Takemura, S. Y., Ting, C. Y., Huang, S., Lu, Z., Luan, H., Rister, J., Thum, A. S., Yang, M., Hong, S. T., et al. (2008). The neural substrate of spectral preference in Drosophila. Neuron 60, 328–342.

Gegear, R. J., Casselman, A., Waddell, S., and Reppert, S. M. (2008). Cryptochrome mediates light-dependent magnetosensitivity in Drosophila. Nature 454, 1014–1018.

Giepmans, B. N., Adams, S. R., Ellisman, M. H., and Tsien, R. Y. (2006). The fluorescent toolbox for assessing protein location and function. Science (New York, N.Y.) 312, 217–224.

Golic, K. G. (1991). Site-specific recombination between homologous chromosomes in Drosophila. Science (New York, N.Y.) 252, 958–961.

Golic, K. G., and Lindquist, S. (1989). The FLP recombinase of yeast catalyzes site-specific recombination in the Drosophila genome. Cell 59, 499–509.

Gorcyzca, M., and Hall, J. C. (1987). The INSECTAVOX, an integrated device for recording and amplifying courtship songs. Drosoph. Inf. Serv. 157–160.

Gordon, M., and Scott, K. (2009a). A motor neuron involved in taste behavior? Neuron.

Gordon, M. D., and Scott, K. (2009b). Motor control in a Drosophila taste circuit. Neuron 61, 373–384.

Gradinaru, V., Thompson, K. R., and Deisseroth, K. (2008). eNpHR: A Natronomonas halorhodopsin enhanced for optogenetic applications. Brain Cell Biol. 36, 129–139.

Gray, J. M., Hill, J. J., and Bargmann, C. I. (2005). A circuit for navigation in Caenorhabditis elegans. *Proc. Natl. Acad. Sci. USA* **102**, 3184–3191.

Greenspan, R. J. (1997). A kinder, gentler genetic analysis of behavior: Dissection gives way to modulation. *Curr. Opin. Neurobiol.* **7**, 805–811.

Greenspan, R. J., and Ferveur, J. F. (2000). Courtship in *Drosophila. Annu. Rev. Genet.* **34**, 205–232.

Griffith, L. C., Verselis, L. M., Aitken, K. M., Kyriacou, C. P., Danho, W., and Greenspan, R. J. (1993). Inhibition of calcium/calmodulin-dependent protein kinase in *Drosophila* disrupts behavioral plasticity. *Neuron* **10**, 501–509.

Groth, A. C., Fish, M., Nusse, R., and Calos, M. P. (2004). Construction of transgenic *Drosophila* by using the site-specific integrase from phage phiC31. *Genetics* **166**, 1775–1782.

Gu, H., and O'Dowd, D. K. (2006). Cholinergic synaptic transmission in adult *Drosophila* Kenyon cells in situ. *J. Neurosci.* **26**, 265–272.

Gu, H., and O'Dowd, D. K. (2007). Whole cell recordings from brain of adult *Drosophila. J. Vis. Exp.* 248.

Guerrero, G., Siegel, M. S., Roska, B., Loots, E., and Isacoff, E. Y. (2002). Tuning FlaSh: Redesign of the dynamics, voltage range, and color of the genetically encoded optical sensor of membrane potential. *Biophys. J.* **83**, 3607–3618.

Guerrero, G., Reiff, D. F., Agarwal, G., Ball, R. W., Borst, A., Goodman, C. S., and Isacoff, E. Y. (2005). Heterogeneity in synaptic transmission along a *Drosophila* larval motor axon. *Nat. Neurosci.* **8**, 1188–1196.

Guss, K. A., Nelson, C. E., Hudson, A., Kraus, M. E., and Carroll, S. B. (2001). Control of a genetic regulatory network by a selector gene. *Science (New York, N.Y.)* **292**, 1164–1167.

Hall, J. C. (1979). Control of male reproductive behavior by the central nervous system of *Drosophila*: Dissection of a courtship pathway by genetic mosaics. *Genetics* **92**, 437–457.

Hallem, E. A., and Carlson, J. R. (2006). Coding of odors by a receptor repertoire. *Cell* **125**, 143–160.

Hamada, F. N., Rosenzweig, M., Kang, K., Pulver, S. R., Ghezzi, A., Jegla, T. J., and Garrity, P. A. (2008). An internal thermal sensor controlling temperature preference in *Drosophila. Nature* **454**, 217–220.

Hammond, S., and O'Shea, M. (2007). Escape flight initiation in the fly. *J. Comp. Physiol.* **193**, 471–476.

Han, P. L., Meller, V., and Davis, R. L. (1996). The *Drosophila* brain revisited by enhancer detection. *J. Neurobiol.* **31**, 88–102.

Hardie, R. C. (1991). Voltage-sensitive potassium channels in *Drosophila* photoreceptors. *J. Neurosci.* **11**, 3079–3095.

Hasemeyer, M., Yapici, N., Heberlein, U., and Dickson, B. J. (2009). Sensory neurons in the *Drosophila* genital tract regulate female reproductive behavior. *Neuron* **61**, 511–518.

Hayashi, S., Ito, K., Sado, Y., Taniguchi, M., Akimoto, A., Takeuchi, H., Aigaki, T., Matsuzaki, F., Nakagoshi, H., Tanimura, T., et al. (2002). GETDB, a database compiling expression patterns and molecular locations of a collection of Gal4 enhancer traps. *Genesis* **34**, 58–61.

Heidmann, D., and Lehner, C. F. (2001). Reduction of Cre recombinase toxicity in proliferating *Drosophila* cells by estrogen-dependent activity regulation. *Dev. Genes Evol.* **211**, 458–465.

Heisenberg, M., Heusipp, M., and Wanke, C. (1995). Structural plasticity in the *Drosophila* brain. *J. Neurosci.* **15**, 1951–1960.

Hendel, T., Mank, M., Schnell, B., Griesbeck, O., Borst, A., and Reiff, D. F. (2008). Fluorescence changes of genetic calcium indicators and OGB-1 correlated with neural activity and calcium in vivo and in vitro. *J. Neurosci.* **28**, 7399–7411.

Hidalgo, A., and Brand, A. H. (1997). Targeted neuronal ablation: The role of pioneer neurons in guidance and fasciculation in the CNS of *Drosophila. Development (Cambridge, England)* **124**, 3253–3262.

Hidalgo, A., Urban, J., and Brand, A. H. (1995). Targeted ablation of glia disrupts axon tract formation in the Drosophila CNS. Development (Cambridge, England) 121, 3703–3712.

Hiesinger, P. R., Zhai, R. G., Zhou, Y., Koh, T. W., Mehta, S. Q., Schulze, K. L., Cao, Y., Verstreken, P., Clandinin, T. R., Fischbach, K. F., et al. (2006). Activity-independent prespecification of synaptic partners in the visual map of Drosophila. Curr. Biol. 16, 1835–1843.

Hinner, M. J., Hubener, G., and Fromherz, P. (2006). Genetic targeting of individual cells with a voltage-sensitive dye through enzymatic activation of membrane binding. Chembiochem 7, 495–505.

Hinshaw, J. E. (2000). Dynamin and its role in membrane fission. Annu. Rev. Cell Dev. Biol. 16, 483–519.

Hires, S. A., Tian, L., and Looger, L. L. (2008a). Reporting neural activity with genetically encoded calcium indicators. Brain Cell Biol. 36, 69–86.

Hires, S. A., Zhu, Y., and Tsien, R. Y. (2008b). Optical measurement of synaptic glutamate spillover and reuptake by linker optimized glutamate-sensitive fluorescent reporters. Proc. Natl. Acad. Sci. USA 105, 4411–4416.

Hiromi, Y., Kuroiwa, A., and Gehring, W. J. (1985). Control elements of the Drosophila segmentation gene fushi tarazu. Cell 43, 603–613.

Hodge, J. J., Choi, J. C., O'Kane, C. J., and Griffith, L. C. (2005). Shaw potassium channel genes in Drosophila. J. Neurobiol. 63, 235–254.

Holmes, T. C., Sheeba, V., Mizrak, D., Rubovsky, B., and Dahdal, D. (2007). Circuit-breaking and behavioral analysis by molecular genetic manipulation of neural activity in Drosophila. In "Invertebrate Neurobiology" (G. North and R. Greenspan, eds.), pp. 19–52. Cold Spring Harbor Laboratory Press, Cold Spring Harbor, NY.

Homyk, T., Jr., Szidonya, J., and Suzuki, D. T. (1980). Behavioral mutants of Drosophila melanogaster. III. Isolation and mapping of mutations by direct visual observations of behavioral phenotypes. Mol. Gen. Genet. 177, 553–565.

Hong, S. T., Bang, S., Paik, D., Kang, J., Hwang, S., Jeon, K., Chun, B., Hyun, S., Lee, Y., and Kim, J. (2006). Histamine and its receptors modulate temperature-preference behaviors in Drosophila. J. Neurosci. 26, 7245–7256.

Horn, C., and Handler, A. M. (2005). Site-specific genomic targeting in Drosophila. Proc. Natl. Acad. Sci. USA 102, 12483–12488.

Hotta, Y., and Benzer, S. (1969). Abnormal electroretinograms in visual mutants of Drosophila. Nature 222, 354–356.

Hotta, Y., and Benzer, S. (1970). Genetic dissection of the Drosophila nervous system by means of mosaics. Proc. Natl. Acad. Sci. USA 67, 1156–1163.

Hoyer, S. C., Eckart, A., Herrel, A., Zars, T., Fischer, S. A., Hardie, S. L., and Heisenberg, M. (2008). Octopamine in male aggression of Drosophila. Curr. Biol. 18, 159–167.

Hughes, C. L., and Thomas, J. B. (2007). A sensory feedback circuit coordinates muscle activity in Drosophila. Mol. Cell Neurosci. 35, 383–396.

Hwang, R. Y., Zhong, L., Xu, Y., Johnson, T., Zhang, F., Deisseroth, K., and Tracey, W. D. (2007). Nociceptive neurons protect Drosophila larvae from parasitoid wasps. Curr. Biol. 17, 2105–2116.

Ibanez-Tallon, I., Wen, H., Miwa, J. M., Xing, J., Tekinay, A. B., Ono, F., Brehm, P., and Heintz, N. (2004). Tethering naturally occurring peptide toxins for cell-autonomous modulation of ion channels and receptors in vivo. Neuron 43, 305–311.

Ilius, M., Wolf, R., and Heisenberg, M. (1994). The central complex of Drosophila melanogaster is involved in flight control: Studies on mutants and mosaics of the gene ellipsoid body open. J. Neurogenet. 9, 189–206.

Imlach, W., and McCabe, B. D. (2009). Electrophysiological methods for recording synaptic potentials from the NMJ of Drosophila larvae. J. Vis. Exp.

Inoshita, T., and Tanimura, T. (2006). Cellular identification of water gustatory receptor neurons and their central projection pattern in *Drosophila*. *Proc. Natl. Acad. Sci. USA* **103**, 1094–1099.

Ito, K., Okada, R., Tanaka, N. K., and Awasaki, T. (2003). Cautionary observations on preparing and interpreting brain images using molecular biology-based staining techniques. *Microsc. Res. Tech.* **62**, 170–186.

Ja, W. W., Carvalho, G. B., Mak, E. M., de la Rosa, N. N., Fang, A. Y., Liong, J. C., Brummel, T., and Benzer, S. (2007). Prandiology of *Drosophila* and the CAFE assay. *Proc. Natl. Acad. Sci. USA* **104**, 8253–8256.

Jan, L. Y., and Jan, Y. N. (1976). Properties of the larval neuromuscular junction in *Drosophila melanogaster*. *J. Physiol.* **262**, 189–214.

Jayaraman, V., and Laurent, G. (2007). Evaluating a genetically encoded optical sensor of neural activity using electrophysiology in intact adult fruit flies. *Front. Neural Circuits* **1**, 3.

Jefferis, G. S., Marin, E. C., Stocker, R. F., and Luo, L. (2001). Target neuron prespecification in the olfactory map of *Drosophila*. *Nature* **414**, 204–208.

Jefferis, G. S., Potter, C. J., Chan, A. M., Marin, E. C., Rohlfing, T., Maurer, C. R., Jr., and Luo, L. (2007). Comprehensive maps of *Drosophila* higher olfactory centers: Spatially segregated fruit and pheromone representation. *Cell* **128**, 1187–1203.

Jenett, A., Schindelin, J. E., and Heisenberg, M. (2006). The Virtual Insect Brain protocol: Creating and comparing standardized neuroanatomy. *BMC Bioinformatics* **7**, 544.

Joesch, M., Plett, J., Borst, A., and Reiff, D. F. (2008). Response properties of motion-sensitive visual interneurons in the lobula plate of *Drosophila* melanogaster. *Curr. Biol.* **18**, 368–374.

Joiner, M. A., and Griffith, L. C. (1999). Mapping of the anatomical circuit of CaM kinase-dependent courtship conditioning in *Drosophila*. *Learn. Mem. (Cold Spring Harbor, N.Y.)* **6**, 177–192.

Joiner, M. A., Asztalos, Z., Jones, C. J., Tully, T., and Wu, C. F. (2007). Effects of mutant *Drosophila* K + channel subunits on habituation of the olfactory jump response. *J. Neurogenet.* **21**, 45–58.

Juusola, M., and Hardie, R. C. (2001). Light adaptation in *Drosophila* photoreceptors: I. Response dynamics and signaling efficiency at 25 degrees C. *J. General Physiol.* **117**, 3–25.

Kamikouchi, A., Shimada, T., and Ito, K. (2006). Comprehensive classification of the auditory sensory projections in the brain of the fruit fly *Drosophila* melanogaster. *J. Comp. Neurol.* **499**, 317–356.

Kamikouchi, A., Inagaki, H. K., Effertz, T., Hendrich, O., Fiala, A., Gopfert, M. C., and Ito, K. (2009). The neural basis of *Drosophila* gravity-sensing and hearing. *Nature* **458**, 165–171.

Kaneko, M., Park, J. H., Cheng, Y., Hardin, P. E., and Hall, J. C. (2000). Disruption of synaptic transmission or clock-gene-product oscillations in circadian pacemaker cells of *Drosophila* cause abnormal behavioral rhythms. *J. Neurobiol.* **43**, 207–233.

Karpova, A. Y., Tervo, D. G., Gray, N. W., and Svoboda, K. (2005). Rapid and reversible chemical inactivation of synaptic transmission in genetically targeted neurons. *Neuron* **48**, 727–735.

Katsov, A. Y., and Clandinin, T. R. (2008). Motion processing streams in *Drosophila* are behaviorally specialized. *Neuron* **59**, 322–335.

Kawasaki, F., Felling, R., and Ordway, R. W. (2000). A temperature-sensitive paralytic mutant defines a primary synaptic calcium channel in *Drosophila*. *J. Neurosci.* **20**, 4885–4889.

Kawasaki, F., Zou, B., Xu, X., and Ordway, R. W. (2004). Active zone localization of presynaptic calcium channels encoded by the cacophony locus of *Drosophila*. *J. Neurosci.* **24**, 282–285.

Keene, A. C., and Waddell, S. (2007). *Drosophila* olfactory memory: Single genes to complex neural circuits. *Nat. Rev.* **8**, 341–354.

Kelly, L. E., and Suzuki, D. T. (1974). The effects of increased temperature on electroretinograms of temperature-sensitive paralysis mutants of *Drosophila* melanogaster. *Proc. Natl. Acad. Sci. USA* **71**, 4906–4909.

Kernan, M. J. (2007). Mechanotransduction and auditory transduction in *Drosophila*. *Pflugers Arch.* **454**, 703–720.

Kernan, M., Cowan, D., and Zuker, C. (1994). Genetic dissection of mechanosensory transduction: Mechanoreception-defective mutations of *Drosophila*. *Neuron* **12**, 1195–1206.

Kerr, J. N., and Denk, W. (2008). Imaging in vivo: Watching the brain in action. *Nat. Rev.* **9**, 195–205.

Kerr, R., Lev-Ram, V., Baird, G., Vincent, P., Tsien, R. Y., and Schafer, W. R. (2000). Optical imaging of calcium transients in neurons and pharyngeal muscle of C. elegans. *Neuron* **26**, 583–594.

Kiger, J. A., Jr., Eklund, J. L., Younger, S. H., and O'Kane, C. J. (1999). Transgenic inhibitors identify two roles for protein kinase A in *Drosophila* development. *Genetics* **152**, 281–290.

Kimura, K., Ote, M., Tazawa, T., and Yamamoto, D. (2005). Fruitless specifies sexually dimorphic neural circuitry in the *Drosophila* brain. *Nature* **438**, 229–233.

Kimura, K., Hachiya, T., Koganezawa, M., Tazawa, T., and Yamamoto, D. (2008). Fruitless and doublesex coordinate to generate male-specific neurons that can initiate courtship. *Neuron* **59**, 759–769.

Kitamoto, T. (2001). Conditional modification of behavior in *Drosophila* by targeted expression of a temperature-sensitive shibire allele in defined neurons. *J. Neurobiol.* **47**, 81–92.

Kitamoto, T. (2002). Conditional disruption of synaptic transmission induces male-male courtship behavior in *Drosophila*. *Proc. Natl. Acad. Sci. USA* **99**, 13232–13237.

Koenig, J. H., and Ikeda, K. (1983). Evidence for a presynaptic blockage of transmission in a temperature-sensitive mutant of Drosophila. *J. Neurobiol.* **14**, 411–419.

Koh, Y. H., Popova, E., Thomas, U., Griffith, L. C., and Budnik, V. (1999). Regulation of DLG localization at synapses by CaMKII-dependent phosphorylation. *Cell* **98**, 353–363.

Konopka, R. J., and Benzer, S. (1971). Clock mutants of *Drosophila* melanogaster. *Proc. Natl. Acad. Sci. USA* **68**, 2112–2116.

Kramer, J. M., and Staveley, B. E. (2003). GAL4 causes developmental defects and apoptosis when expressed in the developing eye of *Drosophila* melanogaster. *Genet. Mol. Res.* **2**, 43–47.

Kreher, S. A., Mathew, D., Kim, J., and Carlson, J. R. (2008). Translation of sensory input into behavioral output via an olfactory system. *Neuron* **59**, 110–124.

Lai, S. L., and Lee, T. (2006). Genetic mosaic with dual binary transcriptional systems in *Drosophila*. *Nat. Neurosci.* **9**, 703–709.

Laissue, P. P., Reiter, C., Hiesinger, P. R., Halter, S., Fischbach, K. F., and Stocker, R. F. (1999). Three-dimensional reconstruction of the antennal lobe in *Drosophila* melanogaster. *J. Comp. Neurol* **405**, 543–552.

Lalli, G., Bohnert, S., Deinhardt, K., Verastegui, C., and Schiavo, G. (2003). The journey of tetanus and botulinum neurotoxins in neurons. *Trends Microbiol.* **11**, 431–437.

Larsen, C. W., Hirst, E., Alexandre, C., and Vincent, J. P. (2003). Segment boundary formation in *Drosophila* embryos. *Development (Cambridge, England)* **130**, 5625–5635.

Lavis, L. D., Chao, T. Y., and Raines, R. T. (2006). Fluorogenic label for biomolecular imaging. *ACS Chem. Biol.* **1**, 252–260.

Leal, S. M., Kumar, N., and Neckameyer, W. S. (2004). GABAergic modulation of motor-driven behaviors in juvenile *Drosophila* and evidence for a nonbehavioral role for GABA transport. *J. Neurobiol.* **61**, 189–208.

Lee, T., and Luo, L. (1999). Mosaic analysis with a repressible cell marker for studies of gene function in neuronal morphogenesis. *Neuron* **22**, 451–461.

Lee, T., and Luo, L. (2001). Mosaic analysis with a repressible cell marker (MARCM) for *Drosophila* neural development. *Trends Neurosci.* **24**, 251–254.

Lehmann, F. O., and Dickinson, M. H. (2001). The production of elevated flight force compromises manoeuvrability in the fruit fly *Drosophila* melanogaster. *J. Exp. Biol.* **204**, 627–635.

Lerchner, W., Xiao, C., Nashmi, R., Slimko, E. M., van Trigt, L., Lester, H. A., and Anderson, D. J. (2007). Reversible silencing of neuronal excitability in behaving mice by a genetically targeted, ivermectin-gated Cl- channel. *Neuron* **54**, 35–49.

Li, W., Ohlmeyer, J. T., Lane, M. E., and Kalderon, D. (1995). Function of protein kinase A in hedgehog signal transduction and *Drosophila* imaginal disc development. *Cell* **80**, 553–562.

Li, X., Gutierrez, D. V., Hanson, M. G., Han, J., Mark, M. D., Chiel, H., Hegemann, P., Landmesser, L. T., and Herlitze, S. (2005). Fast noninvasive activation and inhibition of neural and network activity by vertebrate rhodopsin and green algae channelrhodopsin. *Proc. Natl. Acad. Sci. USA* **102**, 17816–17821.

Lichtman, J. W., Livet, J., and Sanes, J. R. (2008). A technicolour approach to the connectome. *Nat. Rev.* **9**, 417–422.

Lilly, M., and Carlson, J. (1990). Smellblind: A gene required for Drosophila olfaction. *Genetics* **124**, 293–302.

Lima, S. Q., and Miesenbock, G. (2005). Remote control of behavior through genetically targeted photostimulation of neurons. *Cell* **121**, 141–152.

Lin, D. M., Auld, V. J., and Goodman, C. S. (1995). Targeted neuronal cell ablation in the *Drosophila* embryo: Pathfinding by follower growth cones in the absence of pioneers. *Neuron* **14**, 707–715.

Lin, Y. J., Seroude, L., and Benzer, S. (1998). Extended life-span and stress resistance in the *Drosophila* mutant methuselah. *Science (New York, N.Y.)* **282**, 943–946.

Lippincott-Schwartz, J., and Patterson, G. H. (2008). Fluorescent proteins for photoactivation experiments. *Methods Cell Biol.* **85**, 45–61.

Littleton, J. T., and Ganetzky, B. (2000). Ion channels and synaptic organization: Analysis of the *Drosophila* genome. *Neuron* **26**, 35–43.

Liu, L., Wolf, R., Ernst, R., and Heisenberg, M. (1999). Context generalization in *Drosophila* visual learning requires the mushroom bodies. *Nature* **400**, 753–756.

Liu, L., Yermolaieva, O., Johnson, W. A., Abboud, F. M., and Welsh, M. J. (2003). Identification and function of thermosensory neurons in *Drosophila* larvae. *Nat. Neurosci.* **6**, 267–273.

Liu, G., Seiler, H., Wen, A., Zars, T., Ito, K., Wolf, R., Heisenberg, M., and Liu, L. (2006). Distinct memory traces for two visual features in the *Drosophila* brain. *Nature* **439**, 551–556.

Liu, L., Li, Y., Wang, R., Yin, C., Dong, Q., Hing, H., Kim, C., and Welsh, M. J. (2007). *Drosophila* hygrosensation requires the TRP channels water witch and nanchung. *Nature* **450**, 294–298.

Livet, J., Weissman, T. A., Kang, H., Draft, R. W., Lu, J., Bennis, R. A., Sanes, J. R., and Lichtman, J. W. (2007). Transgenic strategies for combinatorial expression of fluorescent proteins in the nervous system. *Nature* **450**, 56–62.

Lnenicka, G. A., Grizzaffi, J., Lee, B., and Rumpal, N. (2006). Ca2+ dynamics along identified synaptic terminals in *Drosophila* larvae. *J. Neurosci.* **26**, 12283–12293.

Louis, M., Huber, T., Benton, R., Sakmar, T. P., and Vosshall, L. B. (2008a). Bilateral olfactory sensory input enhances chemotaxis behavior. *Nat. Neurosci.* **11**, 187–199.

Louis, M., Piccinotti, S., and Vosshall, L. B. (2008b). High-resolution measurement of odor-driven behavior in *Drosophila* larvae. *J. Vis. Exp.*

Luan, H., and White, B. H. (2007). Combinatorial methods for refined neuronal gene targeting. *Curr. Opin. Neurobiol.* **17**, 572–580.

Luan, H., Lemon, W. C., Peabody, N. C., Pohl, J. B., Zelensky, P. K., Wang, D., Nitabach, M. N., Holmes, T. C., and White, B. H. (2006a). Functional dissection of a neuronal network required for cuticle tanning and wing expansion in *Drosophila*. *J. Neurosci.* **26**, 573–584.

Luan, H., Peabody, N. C., Vinson, C. R., and White, B. H. (2006b). Refined spatial manipulation of neuronal function by combinatorial restriction of transgene expression. *Neuron* **52**, 425–436.

Lukacsovich, T., Asztalos, Z., Awano, W., Baba, K., Kondo, S., Niwa, S., and Yamamoto, D. (2001). Dual-tagging gene trap of novel genes in *Drosophila* melanogaster. *Genetics* **157,** 727–742.

Lukacsovich, T., Hamada, N., Miyazaki, S., Kimpara, A., and Yamamoto, D. (2008). A new versatile gene-trap vector for insect transgenics. *Arch. Insect Biochem. Physiol.* **69,** 168–175.

Luo, L., Callaway, E. M., and Svoboda, K. (2008). Genetic dissection of neural circuits. *Neuron* **57,** 634–660.

Ma, J., and Ptashne, M. (1987). The carboxy-terminal 30 amino acids of GAL4 are recognized by GAL80. *Cell* **50,** 137–142.

Maimon, G., Straw, A. D., and Dickinson, M. H. (2008). A simple vision-based algorithm for decision making in flying *Drosophila*. *Curr. Biol.* **18,** 464–470.

Mank, M., Reiff, D. F., Heim, N., Friedrich, M. W., Borst, A., and Griesbeck, O. (2006). A FRET-based calcium biosensor with fast signal kinetics and high fluorescence change. *Biophys. J.* **90,** 1790–1796.

Mank, M., Santos, A. F., Direnberger, S., Mrsic-Flogel, T. D., Hofer, S. B., Stein, V., Hendel, T., Reiff, D. F., Levelt, C., Borst, A., *et al.* (2008). A genetically encoded calcium indicator for chronic *in vivo* two-photon imaging. *Nat. Methods* **5,** 805–811.

Manoli, D. S., and Baker, B. S. (2004). Median bundle neurons coordinate behaviours during *Drosophila* male courtship. *Nature* **430,** 564–569.

Manoli, D. S., Foss, M., Villella, A., Taylor, B. J., Hall, J. C., and Baker, B. S. (2005). Male-specific fruitless specifies the neural substrates of *Drosophila* courtship behaviour. *Nature* **436,** 395–400.

Manseau, L., Baradaran, A., Brower, D., Budhu, A., Elefant, F., Phan, H., Philp, A. V., Yang, M., Glover, D., Kaiser, K., *et al.* (1997). GAL4 enhancer traps expressed in the embryo, larval brain, imaginal discs, and ovary of *Drosophila*. *Dev. Dyn.* **209,** 310–322.

Mao, T., O'Connor, D. H., Scheuss, V., Nakai, J., and Svoboda, K. (2008). Characterization and subcellular targeting of GCaMP-type genetically-encoded calcium indicators. *PLoS ONE* **3,** e1796.

Marder, E., and Bucher, D. (2007). Understanding circuit dynamics using the stomatogastric nervous system of lobsters and crabs. *Annu. Rev. Physiol.* **69,** 291–316.

Marella, S., Fischler, W., Kong, P., Asgarian, S., Rueckert, E., and Scott, K. (2006). Imaging taste responses in the fly brain reveals a functional map of taste category and behavior. *Neuron* **49,** 285–295.

Marin, E. C., Jefferis, G. S., Komiyama, T., Zhu, H., and Luo, L. (2002). Representation of the glomerular olfactory map in the *Drosophila* brain. *Cell* **109,** 243–255.

Markstein, M., Zinzen, R., Markstein, P., Yee, K. P., Erives, A., Stathopoulos, A., and Levine, M. (2004). A regulatory code for neurogenic gene expression in the *Drosophila* embryo. *Development (Cambridge, England)* **131,** 2387–2394.

Martin, J. R. (2003). Locomotor activity: A complex behavioural trait to unravel. *Behav. Processes* **64,** 145–160.

Martin, J. R. (2008). *In vivo* brain imaging: Fluorescence or bioluminescence, which to choose? *J. Neurogenet.* 1–23.

Martin, J. R., Keller, A., and Sweeney, S. T. (2002). Targeted expression of tetanus toxin: A new tool to study the neurobiology of behavior. *Adv. Genet.* **47,** 1–47.

Martin, J. R., Rogers, K. L., Chagneau, C., and Brulet, P. (2007). *In vivo* bioluminescence imaging of Ca signalling in the brain of *Drosophila*. *PLoS ONE* **2,** e275.

Mathey-Prevot, B., and Perrimon, N. (2006). *Drosophila* genome-wide RNAi screens: Are they delivering the promise? *Cold Spring Harb. Symp. Quant. Biol.* **71,** 141–148.

Matthies, H. J., and Broadie, K. (2003). Techniques to dissect cellular and subcellular function in the *Drosophila* nervous system. *Methods Cell Biol.* **71,** 195–265.

McGuire, S. E., Le, P. T., Osborn, A. J., Matsumoto, K., and Davis, R. L. (2003). Spatiotemporal rescue of memory dysfunction in *Drosophila*. *Science (New York, N.Y.)* **302,** 1765–1768.

McGuire, S. E., Roman, G., and Davis, R. L. (2004). Gene expression systems in *Drosophila*: A synthesis of time and space. *Trends Genet.* **20**, 384–391.

McKenna, M., Monte, P., Helfand, S. L., Woodard, C., and Carlson, J. (1989). A simple chemosensory response in *Drosophila* and the isolation of acj mutants in which it is affected. *Proc. Natl. Acad. Sci. USA* **86**, 8118–8122.

McNabb, S. L., Baker, J. D., Agapite, J., Steller, H., Riddiford, L. M., and Truman, J. W. (1997). Disruption of a behavioral sequence by targeted death of peptidergic neurons in *Drosophila*. *Neuron* **19**, 813–823.

Mee, C. J., Pym, E. C., Moffat, K. G., and Baines, R. A. (2004). Regulation of neuronal excitability through pumilio-dependent control of a sodium channel gene. *J. Neurosci.* **24**, 8695–8703.

Melcher, C., and Pankratz, M. J. (2005). Candidate gustatory interneurons modulating feeding behavior in the *Drosophila* brain. *PLoS Biol.* **3**, e305.

Merkle, T., and Wehner, R. (2008). Landmark guidance and vector navigation in outbound desert ants. *J. Exp. Biol.* **211**, 3370–3377.

Mery, F., and Kawecki, T. J. (2002). Experimental evolution of learning ability in fruit flies. *Proc. Natl. Acad. Sci. USA* **99**, 14274–14279.

Miesenbock, G., and Kevrekidis, I. G. (2005). Optical imaging and control of genetically designated neurons in functioning circuits. *Annu. Rev. Neurosci.* **28**, 533–563.

Miesenbock, G., De Angelis, D. A., and Rothman, J. E. (1998). Visualizing secretion and synaptic transmission with pH-sensitive green fluorescent proteins. *Nature* **394**, 192–195.

Miyawaki, A., Llopis, J., Heim, R., McCaffery, J. M., Adams, J. A., Ikura, M., and Tsien, R. Y. (1997). Fluorescent indicators for Ca^{2+} based on green fluorescent proteins and calmodulin. *Nature* **388**, 882–887.

Miyawaki, A., Griesbeck, O., Heim, R., and Tsien, R. Y. (1999). Dynamic and quantitative Ca2+ measurements using improved cameleons. *Proc. Natl. Acad. Sci. USA* **96**, 2135–2140.

Miyawaki, A., Nagai, T., and Mizuno, H. (2005). Engineering fluorescent proteins. *Adv. Biochem. Eng. Biotechnol.* **95**, 1–15.

Moffat, K. G., Gould, J. H., Smith, H. K., and O'Kane, C. J. (1992). Inducible cell ablation in *Drosophila* by cold-sensitive ricin A chain. *Development (Cambridge, England)* **114**, 681–687.

Moline, M. M., Southern, C., and Bejsovec, A. (1999). Directionality of wingless protein transport influences epidermal patterning in the *Drosophila* embryo. *Development (Cambridge, England)* **126**, 4375–4384.

Mondal, K., Dastidar, A. G., Singh, G., Madhusudhanan, S., Gande, S. L., VijayRaghavan, K., and Varadarajan, R. (2007). Design and isolation of temperature-sensitive mutants of Gal4 in yeast and *Drosophila*. *J. Mol. Biol.* **370**, 939–950.

Mongeau, R., Miller, G. A., Chiang, E., and Anderson, D. J. (2003). Neural correlates of competing fear behaviors evoked by an innately aversive stimulus. *J. Neurosci.* **23**, 3855–3868.

Moore, M. S., DeZazzo, J., Luk, A. Y., Tully, T., Singh, C. M., and Heberlein, U. (1998). Ethanol intoxication in *Drosophila*: Genetic and pharmacological evidence for regulation by the cAMP signaling pathway. *Cell* **93**, 997–1007.

Morante, J., and Desplan, C. (2008). The color-vision circuit in the medulla of *Drosophila*. *Curr. Biol.* **18**, 553–565.

Morin, X., Daneman, R., Zavortink, M., and Chia, W. (2001). A protein trap strategy to detect GFP-tagged proteins expressed from their endogenous loci in *Drosophila*. *Proc. Natl. Acad. Sci. USA* **98**, 15050–15055.

Mosca, T. J., Carrillo, R. A., White, B. H., and Keshishian, H. (2005). Dissection of synaptic excitability phenotypes by using a dominant-negative Shaker K+ channel subunit. *Proc. Natl. Acad. Sci. USA* **102**, 3477–3482.

Moses, K., and Rubin, G. M. (1991). Glass encodes a site-specific DNA-binding protein that is regulated in response to positional signals in the developing *Drosophila* eye. *Genes Dev.* **5**, 583–593.

Mundiyanapurath, S., Certel, S., and Kravitz, E. A. (2007). Studying aggression in *Drosophila* (fruit flies). *J. Vis. Exp.* 155.

Murata, Y., Iwasaki, H., Sasaki, M., Inaba, K., and Okamura, Y. (2005). Phosphoinositide phosphatase activity coupled to an intrinsic voltage sensor. *Nature* **435**, 1239–1243.

Murthy, M., Fiete, I., and Laurent, G. (2008). Testing odor response stereotypy in the *Drosophila* mushroom body. *Neuron* **59**, 1009–1023.

Nagai, T., Sawano, A., Park, E. S., and Miyawaki, A. (2001). Circularly permuted green fluorescent proteins engineered to sense Ca^{2+}. *Proc. Natl. Acad. Sci. USA* **98**, 3197–3202.

Nagai, T., Yamada, S., Tominaga, T., Ichikawa, M., and Miyawaki, A. (2004). Expanded dynamic range of fluorescent indicators for Ca(2+) by circularly permuted yellow fluorescent proteins. *Proc. Natl. Acad. Sci. USA* **101**, 10554–10559.

Nagel, G., Szellas, T., Huhn, W., Kateriya, S., Adeishvili, N., Berthold, P., Ollig, D., Hegemann, P., and Bamberg, E. (2003). Channelrhodopsin-2, a directly light-gated cation-selective membrane channel. *Proc. Natl. Acad. Sci. USA* **100**, 13940–13945.

Nagel, G., Brauner, M., Liewald, J. F., Adeishvili, N., Bamberg, E., and Gottschalk, A. (2005). Light activation of channelrhodopsin-2 in excitable cells of Caenorhabditis elegans triggers rapid behavioral responses. *Curr. Biol.* **15**, 2279–2284.

Nakai, J., Ohkura, M., and Imoto, K. (2001). A high signal-to-noise Ca(2+) probe composed of a single green fluorescent protein. *Nat. Biotechnol.* **19**, 137–141.

Nassel, D. R., and Homberg, U. (2006). Neuropeptides in interneurons of the insect brain. *Cell Tissue Res.* **326**, 1–24.

Neuser, K., Triphan, T., Mronz, M., Poeck, B., and Strauss, R. (2008). Analysis of a spatial orientation memory in *Drosophila*. *Nature* **453**, 1244–1247.

Newsome, T. P., Asling, B., and Dickson, B. J. (2000). Analysis of *Drosophila* photoreceptor axon guidance in eye-specific mosaics. *Development (Cambridge, England)* **127**, 851–860.

Ng, M., Roorda, R. D., Lima, S. Q., Zemelman, B. V., Morcillo, P., and Miesenbock, G. (2002). Transmission of olfactory information between three populations of neurons in the antennal lobe of the fly. *Neuron* **36**, 463–474.

Ni, J. Q., Markstein, M., Binari, R., Pfeiffer, B., Liu, L. P., Villalta, C., Booker, M., Perkins, L., and Perrimon, N. (2008). Vector and parameters for targeted transgenic RNA interference in *Drosophila* melanogaster. *Nat. Methods* **5**, 49–51.

Nitabach, M. N., Blau, J., and Holmes, T. C. (2002). Electrical silencing of *Drosophila* pacemaker neurons stops the free- running circadian clock. *Cell* **109**, 485–495.

Nitabach, M. N., and Taghert, P. H. (2008). Organization of the *Drosophila* circadian control circuit. *Curr. Biol.* **18**, R84–93.

Nitabach, M. N., Wu, Y., Sheeba, V., Lemon, W. C., Strumbos, J., Zelensky, P. K., White, B. H., and Holmes, T. C. (2006). Electrical hyperexcitation of lateral ventral pacemaker neurons desynchronizes downstream circadian oscillators in the fly circadian circuit and induces multiple behavioral periods. *J. Neurosci.* **26**, 479–489.

Niven, J. E., Vahasoyrinki, M., Kauranen, M., Hardie, R. C., Juusola, M., and Weckstrom, M. (2003). The contribution of Shaker K+ channels to the information capacity of *Drosophila* photoreceptors. *Nature* **421**, 630–634.

O'Dell, K. M. (2003). The voyeurs' guide to *Drosophila* melanogaster courtship. *Behav. Processes* **64**, 211–223.

O'Dowd, D. K. (1995). Voltage-gated currents and firing properties of embryonic *Drosophila* neurons grown in a chemically defined medium. *J. Neurobiol.* **27**, 113–126.

O'Dowd, D. K., and Aldrich, R. W. (1988). Voltage-clamp analysis of sodium channels in wild-type and mutant *Drosophila* neurons. *J. Neurosci.* **8**, 3633–3643.

Oberstein, A., Pare, A., Kaplan, L., and Small, S. (2005). Site-specific transgenesis by Cre-mediated recombination in *Drosophila*. *Nat. Methods* **2**, 583–585.

Olsen, S. R., Bhandawat, V., and Wilson, R. I. (2007). Excitatory interactions between olfactory processing channels in the Drosophila antennal lobe. Neuron **54**, 89–103.

Olsen, S. R., and Wilson, R. I. (2008a). Cracking neural circuits in a tiny brain: New approaches for understanding the neural circuitry of Drosophila. Trends Neurosci. **31**, 512–520.

Olsen, S. R., and Wilson, R. I. (2008b). Lateral presynaptic inhibition mediates gain control in an olfactory circuit. Nature **452**, 956–960.

Orger, M. B., Kampff, A. R., Severi, K. E., Bollmann, J. H., and Engert, F. (2008). Control of visually guided behavior by distinct populations of spinal projection neurons. Nat. Neurosci. **11**, 327–333.

Osterwalder, T., Yoon, K. S., White, B. H., and Keshishian, H. (2001). A conditional tissue-specific transgene expression system using inducible GAL4. Proc. Natl. Acad. Sci. USA **98**, 12596–12601.

Palmer, A. E., Giacomello, M., Kortemme, T., Hires, S. A., Lev-Ram, V., Baker, D., and Tsien, R. Y. (2006). Ca^{2+} indicators based on computationally redesigned calmodulin-peptide pairs. Chem. Biol. **13**, 521–530.

Paradis, S., Sweeney, S. T., and Davis, G. W. (2001). Homeostatic control of presynaptic release is triggered by postsynaptic membrane depolarization. Neuron **30**, 737–749.

Pavlidis, P., and Tanouye, M. A. (1995). Seizures and failures in the giant fiber pathway of Drosophila bang- sensitive paralytic mutants. J. Neurosci. **15**, 5810–5819.

Peng, Y., Xi, W., Zhang, W., Zhang, K., and Guo, A. (2007). Experience improves feature extraction in Drosophila. J. Neurosci. **27**, 5139–5145.

Peabody, N. C., Pohl, J. B., Diao, F., Vreede, A. P., Sandstrom, D. J., Wang, H., Zelensky, P. K., and White, B. H. (2009). Characterization of the decision network for wing expansion in Drosophila using targeted expression of the TRPM8 channel. J. Neurosci. **29**, 3343–3353.

Perazzona, B., Isabel, G., Preat, T., and Davis, R. L. (2004). The role of cAMP response element-binding protein in Drosophila long-term memory. J. Neurosci. **24**, 8823–8828.

Petreanu, L., Huber, D., Sobczyk, A., and Svoboda, K. (2007). Channelrhodopsin-2-assisted circuit mapping of long-range callosal projections. Nat. Neurosci. **10**, 663–668.

Pfeiffer, B. D., Jenett, A., Hammonds, A. S., Ngo, T. T., Misra, S., Murphy, C., Scully, A., Carlson, J. W., Wan, K. H., Laverty, T. R., et al. (2008). Tools for neuroanatomy and neurogenetics in Drosophila. Proc. Natl. Acad. Sci. USA **105**, 9715–9720.

Phelps, C. B., and Brand, A. H. (1998). Ectopic gene expression in Drosophila using GAL4 system. Methods **14**, 367–379.

Phillis, R. W., Bramlage, A. T., Wotus, C., Whittaker, A., Gramates, L. S., Seppala, D., Farahanchi, F., Caruccio, P., and Murphey, R. K. (1993). Isolation of mutations affecting neural circuitry required for grooming behavior in Drosophila melanogaster. Genetics **133**, 581–592.

Pick, S., and Strauss, R. (2005). Goal-driven behavioral adaptations in gap-climbing Drosophila. Curr. Biol. **15**, 1473–1478.

Pitman, J. L., McGill, J. J., Keegan, K. P., and Allada, R. (2006). A dynamic role for the mushroom bodies in promoting sleep in Drosophila. Nature **441**, 753–756.

Poeck, B., Triphan, T., Neuser, K., and Strauss, R. (2008). Locomotor control by the central complex in Drosophila-An analysis of the tay bridge mutant. Dev. Neurobiol. **68**, 1046–1058.

Pologruto, T. A., Yasuda, R., and Svoboda, K. (2004). Monitoring neural activity and [Ca^{2+}] with genetically encoded Ca^{2+} indicators. J. Neurosci. **24**, 9572–9579.

Prinz, A. A., Bucher, D., and Marder, E. (2004). Similar network activity from disparate circuit parameters. Nat. Neurosci. **7**, 1345–1352.

Pulver, S. R., Pashkovski, S. L., Hornstein, N. J., Garrity, P. A., and Griffith, L. C. (2009). Temporal dynamics of neuronal activation by Channelrhodopsin-2 and TRPA1 determine behavioral output in Drosophila larvae. J. Neurophysiol.

Pyza, E., and Meinertzhagen, I. A. (1999). Daily rhythmic changes of cell size and shape in the first optic neuropil in Drosophila melanogaster. J. Neurobiol. **40**, 77–88.

Quinones-Coello, A. T., Petrella, L. N., Ayers, K., Melillo, A., Mazzalupo, S., Hudson, A. M., Wang, S., Castiblanco, C., Buszczak, M., Hoskins, R. A., and Cooley, L. (2007). Exploring strategies for protein trapping in Drosophila. Genetics 175, 1089–1104.

Raghu, S. V., Joesch, M., Borst, A., and Reiff, D. F. (2007). Synaptic organization of lobula plate tangential cells in Drosophila: Gamma-aminobutyric acid receptors and chemical release sites. J. Comp. Neurol. 502, 598–610.

Ramsey, I. S., Moran, M. M., Chong, J. A., and Clapham, D. E. (2006). A voltage-gated proton-selective channel lacking the pore domain. Nature 440, 1213–1216.

Reiff, D. F., Ihring, A., Guerrero, G., Isacoff, E. Y., Joesch, M., Nakai, J., and Borst, A. (2005). In vivo performance of genetically encoded indicators of neural activity in flies. J. Neurosci. 25, 4766–4778.

Reijmers, L. G., Perkins, B. L., Matsuo, N., and Mayford, M. (2007). Localization of a stable neural correlate of associative memory. Science (New York, N.Y.) 317, 1230–1233.

Rein, K., Zockler, M., Mader, M. T., Grubel, C., and Heisenberg, M. (2002). The Drosophila standard brain. Curr. Biol. 12, 227–231.

Reiser, M. B., and Dickinson, M. H. (2008). A modular display system for insect behavioral neuroscience. J. Neurosci. Methods 167, 127–139.

Renn, S. C., Park, J. H., Rosbash, M., Hall, J. C., and Taghert, P. H. (1999). A pdf neuropeptide gene mutation and ablation of PDF neurons each cause severe abnormalities of behavioral circadian rhythms in Drosophila. Cell 99, 791–802.

Riemensperger, T., Voller, T., Stock, P., Buchner, E., and Fiala, A. (2005). Punishment prediction by dopaminergic neurons in Drosophila. Curr. Biol. 15, 1953–1960.

Rister, J., and Heisenberg, M. (2006). Distinct functions of neuronal synaptobrevin in developing and mature fly photoreceptors. J. Neurobiol. 66, 1271–1284.

Rister, J., Pauls, D., Schnell, B., Ting, C. Y., Lee, C. H., Sinakevitch, I., Morante, J., Strausfeld, N. J., Ito, K., and Heisenberg, M. (2007). Dissection of the peripheral motion channel in the visual system of Drosophila melanogaster. Neuron 56, 155–170.

Ritter, D. A., Bhatt, D. H., and Fetcho, J. R. (2001). In vivo imaging of zebrafish reveals differences in the spinal networks for escape and swimming movements. J. Neurosci. 21, 8956–8965.

Ritzenthaler, S., Suzuki, E., and Chiba, A. (2000). Postsynaptic filopodia in muscle cells interact with innervating motoneuron axons. Nat. Neurosci. 3, 1012–1017.

Robertson, K., Mergliano, J., and Minden, J. S. (2003). Dissecting Drosophila embryonic brain development using photoactivated gene expression. Dev. Biol. 260, 124–137.

Robinson, I. M., Ranjan, R., and Schwarz, T. L. (2002). Synaptotagmins I and IV promote transmitter release independently of Ca(2+) binding in the C(2)A domain. Nature 418, 336–340.

Rodan, A. R., Kiger, J. A., Jr., and Heberlein, U. (2002). Functional dissection of neuroanatomical loci regulating ethanol sensitivity in Drosophila. J. Neurosci. 22, 9490–9501.

Rodin, S., and Georgiev, P. (2005). Handling three regulatory elements in one transgene: Combined use of cre-lox, FLP-FRT, and I-SceI recombination systems. BioTechniques 39, 871–876.

Roeder, T. (2005). Tyramine and octopamine: Ruling behavior and metabolism. Annu. Rev. Entomol. 50, 447–477.

Rohrbough, J., and Broadie, K. (2002). Electrophysiological analysis of synaptic transmission in central neurons of Drosophila larvae. J. Neurophysiol. 88, 847–860.

Roman, G., Endo, K., Zong, L., and Davis, R. L. (2001). P[Switch], a system for spatial and temporal control of gene expression in Drosophila melanogaster. Proc. Natl. Acad. Sci. USA 98, 12602–12607.

Rong, Y. S., and Golic, K. G. (2000). Gene targeting by homologous recombination in Drosophila. Science (New York, N.Y.) 288, 2013–2018.

Root, C. M., Semmelhack, J. L., Wong, A. M., Flores, J., and Wang, J. W. (2007). Propagation of olfactory information in Drosophila. Proc. Natl. Acad. Sci. USA 104, 11826–11831.

Rosato, E., and Kyriacou, C. P. (2006). Analysis of locomotor activity rhythms in *Drosophila*. *Nat. Protoc.* **1**, 559–568.

Rosay, P., Armstrong, J. D., Wang, Z., and Kaiser, K. (2001). Synchronized neural activity in the *Drosophila* memory centers and its modulation by amnesiac. *Neuron* **30**, 759–770.

Rosenzweig, M., Brennan, K. M., Tayler, T. D., Phelps, P. O., Patapoutian, A., and Garrity, P. A. (2005). The *Drosophila* ortholog of vertebrate TRPA1 regulates thermotaxis. *Genes Dev.* **19**, 419–424.

Rosenzweig, M., Kang, K., and Garrity, P. A. (2008). Distinct TRP channels are required for warm and cool avoidance in *Drosophila* melanogaster. *Proc. Natl. Acad. Sci. USA* **105**, 14668–14673.

Rubin, G. M., and Spradling, A. C. (1982). Genetic transformation of *Drosophila* with transposable element vectors. *Science (New York, N.Y.)* **218**, 348–353.

Sachse, S., Rueckert, E., Keller, A., Okada, R., Tanaka, N. K., Ito, K., and Vosshall, L. B. (2007). Activity-dependent plasticity in an olfactory circuit. *Neuron* **56**, 838–850.

Saito, M., and Wu, C. F. (1991). Expression of ion channels and mutational effects in giant *Drosophila* neurons differentiated from cell division-arrested embryonic neuroblasts. *J. Neurosci.* **11**, 2135–2150.

Sakai, R., Repunte-Canonigo, V., Raj, C. D., and Knopfel, T. (2001). Design and characterization of a DNA-encoded, voltage-sensitive fluorescent protein. *Eur. J. Neurosci.* **13**, 2314–2318.

Sanchez-Soriano, N., Bottenberg, W., Fiala, A., Haessler, U., Kerassoviti, A., Knust, E., Lohr, R., and Prokop, A. (2005). Are dendrites in *Drosophila* homologous to vertebrate dendrites? *Dev. Biol.* **288**, 126–138.

Sasaki, M., Takagi, M., and Okamura, Y. (2006). A voltage sensor-domain protein is a voltage-gated proton channel. *Science (New York, N.Y.)* **312**, 589–592.

Sayeed, O., and Benzer, S. (1996). Behavioral genetics of thermosensation and hygrosensation in *Drosophila*. *Proc. Natl. Acad. Sci. USA* **93**, 6079–6084.

Schiavo, G., Matteoli, M., and Montecucco, C. (2000). Neurotoxins affecting neuroexocytosis. *Physiol. Rev.* **80**, 717–766.

Schroder-Lang, S., Schwarzel, M., Seifert, R., Strunker, T., Kateriya, S., Looser, J., Watanabe, M., Kaupp, U. B., Hegemann, P., and Nagel, G. (2007). Fast manipulation of cellular cAMP level by light in vivo. *Nat. Methods* **4**, 39–42.

Schroll, C., Riemensperger, T., Bucher, D., Ehmer, J., Voller, T., Erbguth, K., Gerber, B., Hendel, T., Nagel, G., Buchner, E., and Fiala, A. (2006). Light-induced activation of distinct modulatory neurons triggers appetitive or aversive learning in *Drosophila* larvae. *Curr. Biol.* **16**, 1741–1747.

Schulz, D. J., Goaillard, J. M., and Marder, E. (2006). Variable channel expression in identified single and electrically coupled neurons in different animals. *Nat. Neurosci.* **9**, 356–362.

Scott, E. K., Raabe, T., and Luo, L. (2002). Structure of the vertical and horizontal system neurons of the lobula plate in *Drosophila*. *J. Comp. Neurol.* **454**, 470–481.

Seecof, R. L., Alleaume, N., Teplitz, R. L., and Gerson, I. (1971). Differentiation of neurons and myocytes in cell cultures made from *Drosophila* gastrulae. *Exp. Cell Res.* **69**, 161–173.

Sepp, K. J., and Auld, V. J. (1999). Conversion of lacZ Enhancer Trap Lines to GAL4 Lines Using Targeted Transposition in *Drosophila* melanogaster. *Genetics* **151**, 1093–1101.

Sepp, K. J., Hong, P., Lizarraga, S. B., Liu, J. S., Mejia, L. A., Walsh, C. A., and Perrimon, N. (2008). Identification of neural outgrowth genes using genome-wide RNAi. *PLoS Genet.* **4**, e1000111.

Shafer, O. T., Kim, D. J., Dunbar-Yaffe, R., Nikolaev, V. O., Lohse, M. J., and Taghert, P. H. (2008). Widespread receptivity to neuropeptide PDF throughout the neuronal circadian clock network of *Drosophila* revealed by real-time cyclic AMP imaging. *Neuron* **58**, 223–237.

Shaner, N. C., Steinbach, P. A., and Tsien, R. Y. (2005). A guide to choosing fluorescent proteins. *Nat. Methods* **2**, 905–909.

Shang, Y., Claridge-Chang, A., Sjulson, L., Pypaert, M., and Miesenbock, G. (2007). Excitatory local circuits and their implications for olfactory processing in the fly antennal lobe. *Cell* **128**, 601–612.

Shang, Y., Griffith, L. C., and Rosbash, M. (2008). Light-arousal and circadian photoreception circuits intersect at the large PDF cells of the *Drosophila* brain. *Proc. Natl. Acad. Sci. USA* **105**, 19587–19594.

Sharma, P., Asztalos, Z., Ayyub, C., de Bruyne, M., Dornan, A. J., Gomez-Hernandez, A., Keane, J., Killeen, J., Kramer, S., Madhavan, M., et al. (2005). Isogenic autosomes to be applied in optimal screening for novel mutants with viable phenotypes in *Drosophila* melanogaster. *J. Neurogenet.* **19**, 57–85.

Sharma, Y., Cheung, U., Larsen, E. W., and Eberl, D. F. (2002). PPTGAL, a convenient Gal4 P-element vector for testing expression of enhancer fragments in drosophila. *Genesis* **34**, 115–118.

Sheeba, V., Sharma, V. K., Gu, H., Chou, Y. T., O'Dowd, D. K., and Holmes, T. C. (2008). Pigment dispersing factor-dependent and -independent circadian locomotor behavioral rhythms. *J. Neurosci.* **28**, 217–227.

Shiraiwa, T., and Carlson, J. R. (2007). Proboscis extension response (PER) assay in *Drosophila*. *J. Vis. Exp.* 193.

Sicaeros, B., Campusano, J. M., and O'Dowd, D. K. (2007). Primary neuronal cultures from the brains of late stage *Drosophila* pupae. *J. Vis. Exp.* 200.

Siegal, M. L., and Hartl, D. L. (1996). Transgene Coplacement and high efficiency site-specific recombination with the Cre/loxP system in *Drosophila*. *Genetics* **144**, 715–726.

Siegel, R. W., and Hall, J. C. (1979). Conditioned responses in courtship behavior of normal and mutant *Drosophila*. *Proc. Natl. Acad. Sci. USA* **76**, 3430–3434.

Siegel, M. S., and Isacoff, E. Y. (1997). A genetically encoded optical probe of membrane voltage. *Neuron* **19**, 735–741.

Singh, S., and Wu, C. F. (1989). Complete separation of four potassium currents in *Drosophila*. *Neuron* **2**, 1325–1329.

Sjulson, L., and Miesenbock, G. (2008). Rational optimization and imaging in vivo of a genetically encoded optical voltage reporter. *J. Neurosci.* **28**, 5582–5593.

Song, W., Ranjan, R., Dawson-Scully, K., Bronk, P., Marin, L., Seroude, L., Lin, Y., Nie, Z., Atwood, H., Benzer, S., and Zinsmaier, K. (2002). Presynaptic regulation of neurotransmission in *Drosophila* by the g protein-coupled receptor methuselah. *Neuron* **36**, 105.

St Johnston, D. (2002). The art and design of genetic screens: *Drosophila* melanogaster. *Nat. Rev. Genet.* **3**, 176–188.

Stebbins, M. J., Urlinger, S., Byrne, G., Bello, B., Hillen, W., and Yin, J. C. (2001). Tetracycline-inducible systems for *Drosophila*. *Proc. Natl. Acad. Sci. USA* **98**, 10775–10780.

Stockinger, P., Kvitsiani, D., Rotkopf, S., Tirian, L., and Dickson, B. J. (2005). Neural circuitry that governs *Drosophila* male courtship behavior. *Cell* **121**, 795–807.

Stowers, R. S., and Schwarz, T. L. (1999). A genetic method for generating *Drosophila* eyes composed exclusively of mitotic clones of a single genotype. *Genetics* **152**, 1631–1639.

Strauss, R. (1995). A screen for EMS-induced X-linked locomotor mutants in *Drosophila* melanogaster. *J. Neurogenet.* **10**, 53–54.

Strauss, R., and Heisenberg, M. (1993). A higher control center of locomotor behavior in the *Drosophila* brain. *J. Neurosci.* **13**, 1852–1861.

Struhl, G., and Basler, K. (1993). Organizing activity of wingless protein in *Drosophila*. *Cell* **72**, 527–540.

Su, H., and O'Dowd, D. K. (2003). Fast synaptic currents in *Drosophila* mushroom body Kenyon cells are mediated by alpha-bungarotoxin-sensitive nicotinic acetylcholine receptors and picrotoxin-sensitive GABA receptors. *J. Neurosci.* **23**, 9246–9253.

Suh, G. S., Wong, A. M., Hergarden, A. C., Wang, J. W., Simon, A. F., Benzer, S., Axel, R., and Anderson, D. J. (2004). A single population of olfactory sensory neurons mediates an innate avoidance behaviour in Drosophila. *Nature* **431**, 854–859.

Suh, G. S., Ben-Tabou de Leon, S., Tanimoto, H., Fiala, A., Benzer, S., and Anderson, D. J. (2007). Light activation of an innate olfactory avoidance response in *Drosophila*. *Curr. Biol.* **17**, 905–908.

Sullivan, W., Ashburner, M., and Hawley, S. (2000). *Drosophila* Protocols.

Suster, M. L., Seugnet, L., Bate, M., and Sokolowski, M. B. (2004). Refining GAL4-driven transgene expression in *Drosophila* with a GAL80 enhancer-trap. *Genesis* **39**, 240–245.

Suzuki, D. T., Grigliatti, T., and Williamson, R. (1971). Temperature-sensitive mutations in *Drosophila* melanogaster. VII. A mutation (para-ts) causing reversible adult paralysis. *Proc. Natl. Acad. Sci. USA* **68**, 890–893.

Sweeney, S. T., Broadie, K., Keane, J., Niemann, H., and O'Kane, C. J. (1995). Targeted expression of tetanus toxin light chain in *Drosophila* specifically eliminates synaptic transmission and causes behavioral defects. *Neuron* **14**, 341–351.

Szobota, S., Gorostiza, P., Del Bene, F., Wyart, C., Fortin, D. L., Kolstad, K. D., Tulyathan, O., Volgraf, M., Numano, R., Aaron, H. L., *et al.* (2007). Remote control of neuronal activity with a light-gated glutamate receptor. *Neuron* **54**, 535–545.

Taghert, P. H., and Veenstra, J. A. (2003). *Drosophila* neuropeptide signaling. *Adv. Genet.* **49**, 1–65.

Tan, E. M., Yamaguchi, Y., Horwitz, G. D., Gosgnach, S., Lein, E. S., Goulding, M., Albright, T. D., and Callaway, E. M. (2006). Selective and quickly reversible inactivation of mammalian neurons in vivo using the *Drosophila* allatostatin receptor. *Neuron* **51**, 157–170.

Tang, S., Wolf, R., Xu, S., and Heisenberg, M. (2004). Visual pattern recognition in *Drosophila* is invariant for retinal position. *Science (New York, N.Y.)* **305**, 1020–1022.

Tanouye, M. A., and Wyman, R. J. (1980). Motor outputs of giant nerve fiber in *Drosophila*. *J. Neurophysiol.* **44**, 405–421.

Tauber, E., and Eberl, D. F. (2003). Acoustic communication in *Drosophila*. *Behav. Processes* **64**, 197–210.

Terskikh, A., Fradkov, A., Ermakova, G., Zaraisky, A., Tan, P., Kajava, A. V., Zhao, X., Lukyanov, S., Matz, M., Kim, S., *et al.* (2000). "Fluorescent timer": Protein that changes color with time. *Science (New York, N.Y.)* **290**, 1585–1588.

Thomas, J. B., and Wyman, R. J. (1984). Mutations altering synaptic connectivity between identified neurons in *Drosophila*. *J. Neurosci.* **4**, 530–538.

Thum, A. S., Knapek, S., Rister, J., Dierichs-Schmitt, E., Heisenberg, M., and Tanimoto, H. (2006). Differential potencies of effector genes in adult *Drosophila*. *J. Comp. Neurol.* **498**, 194–203.

Tobin, D., Madsen, D., Kahn-Kirby, A., Peckol, E., Moulder, G., Barstead, R., Maricq, A., and Bargmann, C. (2002). Combinatorial expression of TRPV channel proteins defines their sensory functions and subcellular localization in C. elegans neurons. *Neuron* **35**, 307–318.

Tombola, F., Ulbrich, M. H., and Isacoff, E. Y. (2008). The voltage-gated proton channel Hv1 has two pores, each controlled by one voltage sensor. *Neuron* **58**, 546–556.

Tompkins, L., and Hall, J. C. (1983). Identification of brain sites controlling female receptivity in mosaics of *Drosophila* melanogaster. *Genetics* **103**, 179–195.

Tracey, W. D., Jr., Wilson, R. I., Laurent, G., and Benzer, S. (2003). painless, a *Drosophila* gene essential for nociception. *Cell* **113**, 261–273.

Trichas, G., Begbie, J., and Srinivas, S. (2008). Use of the viral 2A peptide for bicistronic expression in transgenic mice. *BMC Biol.* **6**, 40.

Trimarchi, J. R., and Schneiderman, A. M. (1995). Different neural pathways coordinate *Drosophila* flight initiations evoked by visual and olfactory stimuli. *J. Exp. Biol.* **198**, 1099–1104.

Tsien, R. Y. (2005). Building and breeding molecules to spy on cells and tumors. *FEBS Lett.* **579**, 927–932.

Tsutsui, H., Karasawa, S., Okamura, Y., and Miyawaki, A. (2008). Improving membrane voltage measurements using FRET with new fluorescent proteins. *Nat. Methods* **5**, 683–685.

Tully, T., and Quinn, W. G. (1985). Classical conditioning and retention in normal and mutant *Drosophila* melanogaster. *J. Comp. Physiol. [A]* **157**, 263–277.

Turner, G. C., Bazhenov, M., and Laurent, G. (2008). Olfactory representations by *Drosophila* mushroom body neurons. *J. Neurophysiol.* **99**, 734–746.

Turrigiano, G. G. (2008). The self-tuning neuron: Synaptic scaling of excitatory synapses. *Cell* **135**, 422–435.

Venard, R., and Pichon, Y. (1981). Etude electro-antennographique de la reponse peripherique de l'antenne de *Drosophila* melanogaster a des stimulations odorantes. *C. R. Acad. Sci. Paris* 839–842.

Venard, R., and Pichon, Y. (1984). Electrophysiological analysis of the peripheral response to odours in wild type and smell-deficient *olf* C mutant of *Drosophila melanogaster*. *J. Insect Physiol.* **30**, 1–5.

Venken, K. J., and Bellen, H. J. (2005). Emerging technologies for gene manipulation in *Drosophila* melanogaster. *Nat. Rev. Genet.* **6**, 167–178.

Venken, K. J., He, Y., Hoskins, R. A., and Bellen, H. J. (2006). P[acman]: A BAC transgenic platform for targeted insertion of large DNA fragments in D. melanogaster. *Science (New York, N.Y.)* **314**, 1747–1751.

Verkhusha, V. V., Otsuna, H., Awasaki, T., Oda, H., Tsukita, S., and Ito, K. (2001). An enhanced mutant of red fluorescent protein DsRed for double labeling and developmental timer of neural fiber bundle formation. *J. Biol. Chem.* **276**, 29621–29624.

Villalba-Galea, C. A., Sandtner, W., Dimitrov, D., Mutoh, H., Knopfel, T., and Bezanilla, F. (2009). Charge movement of a voltage-sensitive fluorescent protein. *Biophys. J.* **96**, L19–L21.

Villella, A., and Hall, J. C. (2008). Neurogenetics of courtship and mating in *Drosophila*. *Adv. Genet.* **62**, 67–184.

Villella, A., Gailey, D. A., Berwald, B., Ohshima, S., Barnes, P. T., and Hall, J. C. (1997). Extended reproductive roles of the fruitless gene in Drosophila melanogaster revealed by behavioral analysis of new fru mutants. *Genetics* **147**, 1107–1130.

Vosshall, L. B. (2007). Into the mind of a fly. *Nature* **450**, 193–197.

Vosshall, L. B., Wong, A. M., and Axel, R. (2000). An olfactory sensory map in the fly brain. *Cell* **102**, 147–159.

Waddell, S., Armstrong, J. D., Kitamoto, T., Kaiser, K., and Quinn, W. G. (2000). The amnesiac gene product is expressed in two neurons in the *Drosophila* brain that are critical for memory. *Cell* **103**, 805–813.

Wallace, D. J., Borgloh, S. M., Astori, S., Yang, Y., Bausen, M., Kugler, S., Palmer, A. E., Tsien, R. Y., Sprengel, R., Kerr, J. N., *et al.* (2008). Single-spike detection in vitro and *in vivo* with a genetic Ca^{2+} sensor. *Nat. Methods* **5**, 797–804.

Wang, Y., Wright, N. J., Guo, H., Xie, Z., Svoboda, K., Malinow, R., Smith, D. P., and Zhong, Y. (2001). Genetic manipulation of the odor-evoked distributed neural activity in the *Drosophila* mushroom body. *Neuron* **29**, 267–276.

Wang, J. W., Wong, A. M., Flores, J., Vosshall, L. B., and Axel, R. (2003a). Two-photon calcium imaging reveals an odor-evoked map of activity in the fly brain. *Cell* **112**, 271–282.

Wang, Y., Chiang, A. S., Xia, S., Kitamoto, T., Tully, T., and Zhong, Y. (2003b). Blockade of neurotransmission in *Drosophila* mushroom bodies impairs odor attraction, but not repulsion. *Curr. Biol.* **13**, 1900–1904.

Wang, J., Ma, X., Yang, J. S., Zheng, X., Zugates, C. T., Lee, C. H., and Lee, T. (2004a). Transmembrane/juxtamembrane domain-dependent Dscam distribution and function during mushroom body neuronal morphogenesis. *Neuron* **43**, 663–672.

Wang, Z., Singhvi, A., Kong, P., and Scott, K. (2004b). Taste representations in the *Drosophila* brain. *Cell* **117**, 981–991.

Wang, K. H., Majewska, A., Schummers, J., Farley, B., Hu, C., Sur, M., and Tonegawa, S. (2006). *In vivo* two-photon imaging reveals a role of arc in enhancing orientation specificity in visual cortex. *Cell* **126**, 389–402.

Wang, H., Peca, J., Matsuzaki, M., Matsuzaki, K., Noguchi, J., Qiu, L., Wang, D., Zhang, F., Boyden, E., Deisseroth, K., *et al.* (2007). High-speed mapping of synaptic connectivity using photostimulation in Channelrhodopsin-2 transgenic mice. *Proc. Natl. Acad. Sci. USA* **104**, 8143–8148.

Watson, J. T., Ritzmann, R. E., Zill, S. N., and Pollack, A. J. (2002). Control of obstacle climbing in the cockroach, Blaberus discoidalis. I. Kinematics. *J. Comp. Physiol.* **188**, 39–53.

Weiss, K. R., and Kupfermann, I. (1978). The command neuron concept. *Behav. Brain Sci.* 3–39.

White, B. H., Osterwalder, T. P., Yoon, K. S., Joiner, W. J., Whim, M. D., Kaczmarek, L. K., and Keshishian, H. (2001). Targeted attenuation of electrical activity in *Drosophila* using a genetically modified K(+) channel. *Neuron* **31**, 699–711.

Wickersham, I. R., Lyon, D. C., Barnard, R. J., Mori, T., Finke, S., Conzelmann, K. K., Young, J. A., and Callaway, E. M. (2007). Monosynaptic restriction of transsynaptic tracing from single, genetically targeted neurons. *Neuron* **53**, 639–647.

Williams, D. W., Tyrer, M., and Shepherd, D. (2000). Tau and tau reporters disrupt central projections of sensory neurons in *Drosophila*. *J. Comp. Neurol* **428**, 630–640.

Wilson, R. I., and Laurent, G. (2005). Role of GABAergic inhibition in shaping odor-evoked spatiotemporal patterns in the *Drosophila* antennal lobe. *J. Neurosci.* **25**, 9069–9079.

Wilson, R. I., Turner, G. C., and Laurent, G. (2004). Transformation of olfactory representations in the *Drosophila* antennal lobe. *Science (New York, N.Y.)* **303**, 366–370.

Wittlinger, M., Wehner, R., and Wolf, H. (2006). The ant odometer: Stepping on stilts and stumps. *Science (New York, N.Y.)* **312**, 1965–1967.

Wolf, F. W., Rodan, A. R., Tsai, L. T., and Heberlein, U. (2002). High-resolution analysis of ethanol-induced locomotor stimulation in *Drosophila*. *J. Neurosci.* **22**, 11035–11044.

Wong, A. M., Wang, J. W., and Axel, R. (2002). Spatial representation of the glomerular map in the *Drosophila* protocerebrum. *Cell* **109**, 229–241.

Woodard, C., Huang, T., Sun, H., Helfand, S. L., and Carlson, J. (1989). Genetic analysis of olfactory behavior in *Drosophila*: A new screen yields the ota mutants. *Genetics* **123**, 315–326.

Wright, N. J., and Zhong, Y. (1995). Characterization of K+ currents and the cAMP-dependent modulation in cultured *Drosophila* mushroom body neurons identified by lacZ expression. *J. Neurosci.* **15**, 1025–1034.

Wu, C. F., and Haugland, F. N. (1985). Voltage clamp analysis of membrane currents in larval muscle fibers of *Drosophila*: Alteration of potassium currents in Shaker mutants. *J. Neurosci.* **5**, 2626–2640.

Wu, J. S., and Luo, L. (2006). A protocol for dissecting *Drosophila* melanogaster brains for live imaging or immunostaining. *Nat. Protoc.* **1**, 2110–2115.

Wu, C. F., Ganetzky, B., Jan, L. Y., and Jan, Y. N. (1978). A *Drosophila* mutant with a temperature-sensitive block in nerve conduction. *Proc. Natl. Acad. Sci. USA* **75**, 4047–4051.

Wu, Q., Zhao, Z., and Shen, P. (2005). Regulation of aversion to noxious food by *Drosophila* neuropeptide Y- and insulin-like systems. *Nat. Neurosci.* **8**, 1350–1355.

Wu, Y., Cao, G., Pavlicek, B., Luo, X., and Nitabach, M. N. (2008). Phase coupling of a circadian neuropeptide with rest/activity rhythms detected using a membrane-tethered spider toxin. *PLoS Biol.* **6**, e273.

Wulff, P., Goetz, T., Leppa, E., Linden, A. M., Renzi, M., Swinny, J. D., Vekovischeva, O. Y., Sieghart, W., Somogyi, P., Korpi, E. R., et al. (2007). From synapse to behavior: Rapid modulation of defined neuronal types with engineered GABAA receptors. *Nat. Neurosci.* **10**, 923–929.

Xia, S., Miyashita, T., Fu, T. F., Lin, W. Y., Wu, C. L., Pyzocha, L., Lin, I. R., Saitoe, M., Tully, T., and Chiang, A. S. (2005). NMDA receptors mediate olfactory learning and memory in *Drosophila*. *Curr. Biol.* **15**, 603–615.

Xu, T., and Rubin, G. M. (1993). Analysis of genetic mosaics in developing and adult *Drosophila* tissues. *Development (Cambridge, England)* **117**, 1223–1237.

Xu, S. Y., Cang, C. L., Liu, X. F., Peng, Y. Q., Ye, Y. Z., Zhao, Z. Q., and Guo, A. K. (2006). Thermal nociception in adult *Drosophila*: Behavioral characterization and the role of the painless gene. *Genes Brain Behav.* **5**, 602–613.

Yaksi, E., and Friedrich, R. W. (2006). Reconstruction of firing rate changes across neuronal populations by temporally deconvolved Ca^{2+} imaging. *Nat. Methods* **3**, 377–383.

Yamaguchi, S., Wolf, R., Desplan, C., and Heisenberg, M. (2008). Motion vision is independent of color in *Drosophila*. *Proc. Natl. Acad. Sci. USA* **105**, 4910–4915.

Yamamoto, D. (2007). The neural and genetic substrates of sexual behavior in *Drosophila*. *Adv. Genet.* **59**, 39–66.

Yang, T. T., Cheng, L., and Kain, S. R. (1996). Optimized codon usage and chromophore mutations provide enhanced sensitivity with the green fluorescent protein. *Nucleic Acids Res.* **24**, 4592–4593.

Yang, C. H., Belawat, P., Hafen, E., Jan, L. Y., and Jan, Y. N. (2008). *Drosophila* egg-laying site selection as a system to study simple decision-making processes. *Science (New York, N.Y.)* **319**, 1679–1683.

Yang, C. H., Rumpf, S., Xiang, Y., Gordon, M. D., Song, W., Jan, L. Y., and Jan, Y. N. (2009). Control of the postmating behavioral switch in *Drosophila* females by internal sensory neurons. *Neuron* **61**, 519–526.

Yapici, N., Kim, Y. J., Ribeiro, C., and Dickson, B. J. (2008). A receptor that mediates the post-mating switch in *Drosophila* reproductive behaviour. *Nature* **451**, 33–37.

Yasuda, R., Nimchinsky, E. A., Scheuss, V., Pologruto, T. A., Oertner, T. G., Sabatini, B. L., and Svoboda, K. (2004). Imaging calcium concentration dynamics in small neuronal compartments. *Sci STKE* 15.

Yin, J. C., Wallach, J. S., Del Vecchio, M., Wilder, E. L., Zhou, H., Quinn, W. G., and Tully, T. (1994). Induction of a dominant negative CREB transgene specifically blocks long-term memory in *Drosophila*. *Cell* **79**, 49–58.

Yin, J. C., Del Vecchio, M., Zhou, H., and Tully, T. (1995). CREB as a memory modulator: Induced expression of a dCREB2 activator isoform enhances long-term memory in *Drosophila*. *Cell* **81**, 107–115.

Yorozu, S., Wong, A., Fischer, B. J., Dankert, H., Kernan, M. J., Kamikouchi, A., Ito, K., and Anderson, D. J. (2009). Distinct sensory representations of wind and near-field sound in the *Drosophila* brain. *Nature* **458**, 201–205.

Yu, D., Baird, G. S., Tsien, R. Y., and Davis, R. L. (2003). Detection of Calcium Transients in *Drosophila* Mushroom Body Neurons with Camgaroo Reporters. *J. Neurosci.* **23**, 64–72.

Yu, D., Ponomarev, A., and Davis, R. L. (2004). Altered representation of the spatial code for odors after olfactory classical conditioning; memory trace formation by synaptic recruitment. *Neuron* **42**, 437–449.

Yu, D., Keene, A. C., Srivatsan, A., Waddell, S., and Davis, R. L. (2005). *Drosophila* DPM neurons form a delayed and branch-specific memory trace after olfactory classical conditioning. *Cell* **123**, 945–957.

Yu, D., Akalal, D. B., and Davis, R. L. (2006). *Drosophila* alpha/beta mushroom body neurons form a branch-specific, long-term cellular memory trace after spaced olfactory conditioning. *Neuron* **52**, 845–855.

Yurkovic, A., Wang, O., Basu, A. C., and Kravitz, E. A. (2006). Learning and memory associated with aggression in *Drosophila* melanogaster. *Proc. Natl. Acad. Sci. USA* **103**, 17519–17524.

Zars, T. (2001). Two thermosensors in *Drosophila* have different behavioral functions. *J. Comp. Physiol. [A]* **187**, 235–242.

Zars, T., Wolf, R., Davis, R., and Heisenberg, M. (2000). Tissue-specific expression of a type I adenylyl cyclase rescues the rutabaga mutant memory defect: In search of the engram. *Learn. Mem. (Cold Spring Harbor, N.Y.)* **7**, 18–31.

Zeidler, M. P., Tan, C., Bellaiche, Y., Cherry, S., Hader, S., Gayko, U., and Perrimon, N. (2004). Temperature-sensitive control of protein activity by conditionally splicing inteins. *Nat. Biotechnol.* **22**, 871–876.

Zemelman, B. V., Lee, G. A., Ng, M., and Miesenbock, G. (2002). Selective photostimulation of genetically chARGed neurons. *Neuron* **33**, 15–22.

Zemelman, B. V., Nesnas, N., Lee, G. A., and Miesenbock, G. (2003). Photochemical gating of heterologous ion channels: Remote control over genetically designated populations of neurons. *Proc. Natl. Acad. Sci. USA* **100**, 1352–1357.

Zhang, S. D., and Odenwald, W. F. (1995). Misexpression of the white (w) gene triggers male-male courtship in *Drosophila. Proc. Natl. Acad. Sci. USA* **92**, 5525–5529.

Zhang, Y. Q., Rodesch, C. K., and Broadie, K. (2002). Living synaptic vesicle marker: Synaptotagmin-GFP. *Genesis* **34**, 142–145.

Zhang, F., Wang, L. P., Brauner, M., Liewald, J. F., Kay, K., Watzke, N., Wood, P. G., Bamberg, E., Nagel, G., Gottschalk, A., and Deisseroth, K. (2007). Multimodal fast optical interrogation of neural circuitry. *Nature* **446**, 633–639.

Zhang, F., Prigge, M., Beyriere, F., Tsunoda, S. P., Mattis, J., Yizhar, O., Hegemann, P., and Deisseroth, K. (2008a). Red-shifted optogenetic excitation: A tool for fast neural control derived from Volvox carteri. *Nat. Neurosci.* **11**, 631–633.

Zhang, M., Chung, S. H., Fang-Yen, C., Craig, C., Kerr, R. A., Suzuki, H., Samuel, A. D., Mazur, E., and Schafer, W. R. (2008b). A self-regulating feed-forward circuit controlling C. elegans egg-laying behavior. *Curr. Biol.* **18**, 1445–1455.

Zhao, S., Cunha, C., Zhang, F., Liu, Q., Gloss, B., Deisseroth, K., Augustine, G. J., and Feng, G. (2008). Improved expression of halorhodopsin for light-induced silencing of neuronal activity. *Brain Cell Biol.* **36**, 141–154.

Zheng, L., de Polavieja, G. G., Wolfram, V., Asyali, M. H., Hardie, R. C., and Juusola, M. (2006). Feedback network controls photoreceptor output at the layer of first visual synapses in *Drosophila. J. General Physiol.* **127**, 495–510.

Zhong, Y., and Wu, C. F. (1991). Alteration of four identified K^+ currents in Drosophila muscle by mutations in eag. *Science* **252**, 1562–1564.

Zhou, L., Schnitzler, A., Agapite, J., Schwartz, L. M., Steller, H., and Nambu, J. R. (1997). Cooperative functions of the reaper and head involution defective genes in the programmed cell death of *Drosophila* central nervous system midline cells. *Proc. Natl. Acad. Sci. USA* **94**, 5131–5136.

4

A Network of G-Protein Signaling Pathways Control Neuronal Activity in *C. elegans*

Borja Perez-Mansilla and Stephen Nurrish
MRC Cell Biology Unit, MRC Laboratory for Molecular Cell Biology and Department of Neurobiology, Physiology and Pharmacology, University College London, London, United Kingdom

ABSTRACT

The *Caenorhabditis elegans* neuromuscular junction (NMJ) is one of the best studied synapses in any organism. A variety of genetic screens have identified genes required both for the essential steps of neurotransmitter release from

Advances in Genetics, Vol. 65
Copyright 2009, Elsevier Inc. All rights reserved.

0065-2660/09 $35.00
DOI: 10.1016/S0065-2660(09)65004-5

motorneurons as well as the signaling pathways that regulate rates of neurotransmitter release. A number of these regulatory genes encode proteins that converge to regulate neurotransmitter release. In other cases genes are known to regulate signaling at the NMJ but how they act remains unknown.

Many of the proteins that regulate activity at the NMJ participate in a network of heterotrimeric G-protein signaling pathways controlling the release of synaptic vesicles and/or dense-core vesicles (DCVs). At least four heterotrimeric G-proteins ($G\alpha q$, $G\alpha 12$, $G\alpha o$, and $G\alpha s$) act within the motorneurons to control the activity of the NMJ. The $G\alpha q$, $G\alpha 12$, and $G\alpha o$ pathways converge to control production and destruction of the lipid-bound second messenger diacylglycerol (DAG) at sites of neurotransmitter release. DAG acts via at least two effectors, MUNC13 and PKC, to control the release of both neurotransmitters and neuropeptides from motorneurons. The $G\alpha s$ pathway converges with the other three heterotrimeric G-protein pathways downstream of DAG to regulate neuropeptide release. Released neurotransmitters and neuropeptides then act to control contraction of the body-wall muscles to control locomotion.

The lipids and proteins involved in these networks are conserved between *C. elegans* and mammals. Thus, the *C. elegans* NMJ acts as a model synapse to understand how neuronal activity in the human brain is regulated. © 2009, Elsevier Inc.

I. INTRODUCTION

A. Invertebrate neuroscience

Invertebrates have been widely used as models to understand the functioning of the human brain (Sattelle and Buckingham, 2006). Hodgkin and Huxley performed the first voltage-clamp recordings on giant squid axons (Hodgkin, 1964a,b; Hodgkin and Huxley, 1952) and Katz also used squid axons to study synapses (Katz, 1966). Subsequently, the nervous system of a broad range of invertebrates have been examined, these include Aplysia, Ascaris, locusts, leeches, lobsters, crayfish, annelids, and jellyfish. The greatest advances in the understanding of neuronal signaling pathways have come from using the model genetic organisms *Caenorhabditis elegans* and *Drosophila melanogaster*. The *C. elegans* hermaphrodite has 302 neurons and approximately 1000 chemical synapses (Hobert, 2005; White et al., 1986). The anatomy and connectivity of the *C. elegans* nervous system has been extensively mapped from reconstructions of electron micrograph serial sections (White et al., 1986). This makes *C. elegans* the only animal for which we know the neural connectivity of its brain. Despite this simple nervous system a *C. elegans* synapse looks remarkably similar to synapses in other animals (Fig. 4.1) and its nervous system uses many of the same neurotransmitters as mammals and releases these neurotransmitters using the same protein machinery (Brockie and Maricq, 2006; Jorgensen, 2005; Rand, 2007; Richmond, 2005). Genetic approaches in C. *elegans* have identified genes important for neuronal function including those for chemosensation (Bargmann, 2006; Bargmann and Kaplan, 1998; Kaplan, 1996), mechanosensation (Goodman, 2006), thermotaxis (Mori *et al.*, 2007), development of the nervous system (Hobert, 2005), synaptogenesis (Jin, 2005), and learning and memory (Giles *et al.*, 2006). C. *elegans* was the first animal to have its genome completely sequenced (Consortium, 1998) and this allowed for the first time an understanding of the complete set of genes required for a functioning nervous system (Bargmann, 1998). Perhaps the most intensely studied behavior of C. *elegans* is locomotion. As a result we have a detailed, although still incomplete, list of the network of genes that act to regulate signaling at the C. *elegans* neuromuscular junction.

B. *C. elegans* motorneurons

In adult C. *elegans* locomotion involves 95 body-wall muscles whose contraction is controlled by motorneurons. Motorneurons that innervate body-wall muscles (Fig. 4.1) can be divided into two classes: those that release acetylcholine (ACh) (cholinergic) and those that release GABA (GABAergic) (Jorgensen, 2005;

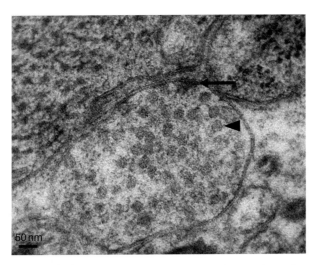

Figure 4.1. An electron micrograph of a section through the body of C. *elegans* shows a neuromuscular junction. The arrow points to the neuromuscular junction and the arrowhead to a synaptic vesicle in the motorneuron. Picture courtesy of Christian Stigloher and Jean-Louis Bessereau, Ecole Normale Superieure, Paris.

Rand, 2007). At body-wall muscles ACh causes muscle contraction whereas GABA causes muscle relaxation. The cholinergic motorneurons are dyadic— they release ACh onto both a muscle and GABAergic motorneuron. Cholinergic motorneurons that cause contraction of ventral muscles also synapse onto GABAergic motorneurons that relax dorsal muscle, and cholinergic motorneurons that cause contraction of dorsal muscles synapse onto GABAergic motorneurons that relax the ventral muscles. Thus, C. elegans moves in a sinusoidal pattern via alternate contraction of the dorsal and ventral muscles and relaxation of the muscles on the opposite side. The synaptic input onto the motorneurons is still unclear, indeed it is not clear that the motorneurons require a synaptic input to release ACh. Interneurons that release glutamate (glutamatergic) do form synapses onto the motorneurons (White et al., 1986). However, animals with defective glutamatergic signaling still move using normal appearing sinusoidal waves (Lee et al., 1999a; Zheng et al., 1999). Alternatively, direct electrical coupling through gap junctions could be important for locomotion. Via electron microscopy gap junctions have been observed both between neurons and muscles (Hall, 2008). Gap junctions could be used to electrically couple cells such that initiation of muscle contraction at one end of the animal triggers a sequential contraction of the remaining muscles. Mutations in the gap junction genes unc-7 and unc-9 cause defects in locomotion, although in both cases forward locomotion is defective whereas reverse locomotion is much less affected (Barnes and Hekimi, 1997; Phelan, 2005). Another suggestion is that stretch receptors within each motorneuron detect the flexing in the body triggered by activation of an adjacent body-wall muscle (Karbowski et al., 2008). This would lead to a sequential activation of the motorneurons along the animal to produce a sinusoidal wave. The cholinergic motorneurons possess long processes much longer than required for them to form synapses onto their postsynaptic muscle target. This extra process could contain stretch receptors and allow motorneurons to detect muscle contraction occurring some defined distance away from the muscle they synapse onto. Finally, the motorneurons could possess intrinsic oscillations in membrane potential that cause muscle contractions as has been proposed in the related motor circuit of Ascaris (Stretton et al., 1992). Interneurons and release of neuromodulators such as neuropeptides could alter the frequency of these oscillations thus controlling the speed and direction of travel. Nonetheless, it is surprising that we still do not know what causes the cholinergic motorneurons to depolarize.

C. The C. elegans NMJ as a model synapse

The C. elegans neuromuscular junction (NMJ) has proved to have many advantages that make it easy to assay its function (Rand, 2007; Rand et al., 1998). There are several reasons why the C. elegans NMJ has been so intensively studied:

– C. *elegans* is the only animal in which the connectivity of the nervous system has been established using reconstruction of serial sections by electron microscopy (EM) (White *et al.*, 1986).
– As in mammals C. *elegans* releases ACh at the NMJ to cause muscle contraction. ACh was the first neurotransmitter identified (Dale, 1914; Loewi, 1921) and the genes required for ACh synthesis and loading into vesicles were some of the very first neuronal proteins identified. A number of groups used this information to identify these ACh-specific genes in C. *elegans* (Johnson *et al.*, 1981; Rand and Russell, 1984).
– Defects in ACh signaling cause defects in behaviors requiring muscle contraction, for example, movement (*unc*oordinated mutants), feeding (*eat* mutants), and egg laying (*egl* mutants). Such mutants were among the first C. *elegans* mutants to be isolated and cloned (Alfonso *et al.*, 1994a; Brenner, 1974). Cloning of these C. *elegans* genes identified genes whose mammalian orthologs turned out to be essential components of the mammalian synaptic release machinery, for example, the C. *elegans* proteins UNC-13 is called MUNC13 (mammalian UNC-13) in mammals (Brose *et al.*, 1995; Maruyama and Brenner, 1991).
– Drugs that specifically alter ACh signaling in C. *elegans* have been extensively used (Rand and Johnson, 1995; Rand and Russell, 1985) and the pharmacology of cholinergic drugs on C. *elegans* is likely to be similar for that in *Ascaris* (Colquhoun *et al.*, 1991; Walker *et al.*, 1992). These consist either of ACh agonists, antagonists, or inhibitors of acetylcholinesterases (AChEs). For example, agonists specific for mammalian G-protein-coupled ACh receptors (muscarinic AChRs) or ACh-gated ion channels (nicotinic receptors) still retain their specificity in C. *elegans*. Exposure to these drugs resulted in easily scored phenotypes allowing genetic screens to identify mutants that altered the response to these drugs and thus ACh signaling (Rand, 2007). Large-scale RNAi screens have also been combined with acetylcholinesterase inhibitor screens to systematically identify genes that regulate ACh signaling (Sieburth *et al.*, 2005; Vashlishan *et al.*, 2008).
– Neuromodulators that alter C. *elegans* behavior, such as serotonin, have been shown to act at least in part by altering ACh release (Nurrish *et al.*, 1999). Genetic screens to identify mutants with an altered response to serotonin have identified signaling pathways that regulate ACh release.
– Finally, recent technical advances have allowed electrophysiological recordings that directly record ACh signaling at body-wall muscles (Richmond, 2006) and methods to depolarize or hyperpolarize selected neurons (Liewald *et al.*, 2008; Nagel *et al.*, 2005; Zhang *et al.*, 2007).

D. Combining genetics with pharmacology

Before describing the signaling pathways controlling ACh signaling at the
C. elegans NMJ we will briefly review the assays that have been most used.

1. Acetylcholinesterase inhibitors

Drugs that alter ACh signaling were described in the very first C. elegans paper
published by Sydney Brenner—lannate and tetramisole (Brenner, 1974). Lannate inhibits acetylcholinesterases, which degrade extracellular ACh, whereas
tetramisole activates ACh-gated ion channels that typically trigger muscle
contraction in response to ACh. Currently, the most commonly used acetylcholinesterase inhibitor is the carbamate aldicarb, although the organophosphate
trichlorfon has also been used. Acetylcholinesterase inhibitors prevent the
hydrolysis of ACh by AChEs. This is predicted to result in the accumulation of
ACh, chronic activation of acetylcholine receptors (AChRs), and hypercontraction of the body-wall muscles. Chronic exposure to 1 mM aldicarb causes death
probably by hypercontraction of the pharyngeal muscle used to eat the bacterial
food source leading to starvation (Hosono and Kamiya, 1991; Miller et al., 1996;
Nguyen et al., 1995). Acute exposure to 1 mM aldicarb over 100 min leads to a
highly reproducible paralysis of wild-type C. elegans that can be reversed if they
are removed to plates lacking aldicarb (Nurrish et al., 1999). Aldicarb alone does
not cause paralysis and death; it is the resulting accumulation of ACh that
poisons the synapse. Mutants that decrease or increase the amount of ACh
released will change the sensitivity of C. elegans to aldicarb. Genetic screens
typically use the chronic aldicarb assay, as assays for living animals against a
background of dead animals is easy to perform, allowing the screening of large
numbers of mutants. Aldicarb screens are not saturated and new aldicarb resistant and hypersensitive mutants are still being identified (Sieburth et al., 2005;
Vashlishan et al., 2008).

2. ACh agonists and antagonists

Other assays use agonists that activate AChRs directly. AChRs can be broadly
divided into two groups. In the first group are the ACh-gated ion channels
referred to as nicotinic AChRs, which are activated in C. elegans by agonists
such as nicotine or levamisole (the active levo-form of tetramisole) (Lewis et al.,
1980a,b; Richmond and Jorgensen, 1999; Waggoner et al., 2000). Studies suggest
that the C. elegans body-wall muscles possess different classes of nicotinic
receptors with different affinities to nicotinic agonists; for a review, see Brown
et al. (2006). Typically, addition of nicotinic agonists is believed to act on the
muscles but nicotinic receptors are widely expressed in muscles and neurons

(Jones and Sattelle, 2004). At least some of the effects of nicotine on *C. elegans* behavior occur via activation of neuronal nicotinic receptors (Feng *et al.*, 2006). The second class of AChRs are the G-protein-coupled AChRs (muscarinic receptors mAChRs) that are coupled to heterotrimeric G-proteins. In *C. elegans* addition of the mAChR agonist arecoline was lethal (Avery, 1993; Brundage *et al.*, 1996) and these effects were blocked by mAChR antagonists such as atropine (Robatzek *et al.*, 2001; You *et al.*, 2006).

II. MAPPING INTRACELLULAR NEURAL NETWORKS WITHIN *C. ELEGANS* MOTORNEURONS

A. The EGL-30(Gαq) pathway

1. DAG stimulates ACh release

A key regulator of neurotransmitter release in both mammalian neurons and *C. elegans* is the membrane-bound second messenger diacylglycerol (DAG) (Lackner *et al.*, 1999; Malenka *et al.*, 1986; Nurrish *et al.*, 1999). The major signaling pathways controlling ACh release in *C. elegans* revolve around the regulated production and destruction of DAG (McMullan and Nurrish, 2007) (Fig. 4.2).

2. EGL-30(Gαq) is a central regulator of motorneuron activity

EGL-30(Gαq) mutants were first identified in a screen for egg-laying (*egl*) mutants that laid eggs slowly and thus retain more eggs within adult animals when compared to wild-type animals (Trent *et al.*, 1983). EGL-30(Gαq) mutations were also recovered in screen for resistance to the lethal effects of either aldicarb or arecoline (Brundage *et al.*, 1996; Miller *et al.*, 1996). These results suggest that EGL-30(Gαq) is required to maintain normal levels of ACh release and to signal downstream of mAChRs. EGL-30(Gαq) is the only Gαq/12 ortholog in *C. elegans* (Bastiani and Mendel, 2006). EGL-30(Gαq) mutations recovered in screens were partial loss-of-function mutations and null mutations in EGL-30(Gαq) are lethal under standard conditions (Brundage *et al.*, 1996). Expression of a constitutively active EGL-30(Gαq) mutant in the cholinergic motorneurons caused animals to move faster than wild type and for animals to become hypersensitive to aldicarb (Lackner *et al.*, 1999). These results suggested that EGL-30(Gαq) acted within the cholinergic motorneurons to increase levels of ACh release. Addition of arecoline to animals caused an increase in DAG at spatially restricted sites within motorneurons and this was blocked by mutations in EGL-30(Gαq) (Lackner *et al.*, 1999). Thus, EGL-30(Gαq) acted downstream

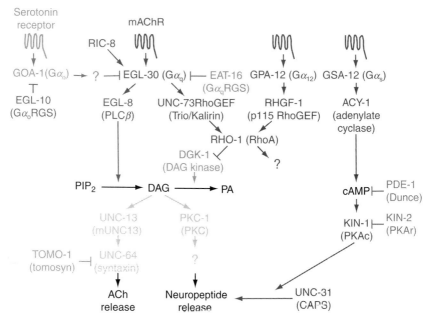

Figure 4.2. A network of heterotrimeric G-proteins regulate acetylcholine (ACh) signaling at the
C. elegans neuromuscular junction to control locomotion. Three pathways, Gαq, Gα12,
and Gαo, act to control the production and destruction of the membrane-bound second
messenger DAG. DAG stimulates the release of both ACh and neuropeptides to control
contraction of the body-wall muscles. The fourth pathway, Gαs pathway, converges
with the other three heterotrimeric G-protein pathways downstream of DAG to regulate
neuropeptide release. Names of C. elegans genes are given first followed by the mammalian ortholog in brackets.

of G-protein-coupled receptors (GPCRs), such as mAChRs, to increase DAG
production within motorneurons and to increase ACh release. In addition to
GPCRs at least two other proteins regulate EGL-30(Gαq) activity. One regulator of EGL-30(Gαq) is RIC-8. Mutations in RIC-8 were first identified in screens
for resistance to aldicarb (Miller et al., 1996). Mutations in RIC-8 were suppressed by gain-of-function mutations in EGL-30(Gαq) suggesting that RIC-8 acted
upstream of EGL-30(Gαq) (Schade et al., 2005). RIC-8 encodes a nonreceptor
activator of Gαq-subunits in both C. elegans and mammals (Miller et al., 2000;
Reynolds et al., 2005; Tall et al., 2003). These results suggested that a significant
amount of EGL-30(Gαq) signaling requires RIC-8. However, it is unclear if
RIC-8 activates EGL-30(Gαq) independently of GPCRs or if RIC-8 prolongs
EGL-30(Gαq) signaling initiated by GPCRs. It is also unclear whether upstream
factors regulate the ability of RIC-8 to activate EGL-30(Gαq). A second

regulator of EGL-30(Gαq) is EAT-16(GαqRGS). Mutations in EAT-16 (GαqRGS) were recovered in screens to suppress gain-of-function mutations in a Gαo-subunit (described later) that caused slow locomotion (Hajdu-Cronin et al., 1999). EAT-16(GαqRGS) contains a regulator of G-protein signaling (RGS) domain that acted as a GTPase promoting (GAP) protein towards EGL-30(Gαq) (Ross and Wilkie, 2000; Siderovski and Willard, 2005). The role of EAT-16(GαqRGS) is to convert EGL-30(Gαq) from an actively signaling GTP-bound form to an inactive nonsignaling GDP-bound form, thus EAT-16(GαqRGS) is a negative regulator of EGL-30(Gαq) signaling. It is unclear whether EAT-16(GαqRGS) is regulated to control EGL-30(Gαq) activity and thus control levels of locomotion and ACh release.

3. EGL-30(Gαq) controls the production of DAG via EGL-8(PLCβ)

In the C. elegans motorneuron the DAG that regulates ACh release is produced by the EGL-8 phospholipase Cβ (PLCβ) (Lackner et al., 1999; Miller et al., 1999), which hydrolyzes phosphatidylinositol-(4,5)-bisphosphate (PIP2) to generate DAG and inositol 1,4,5-triphosphate (IP3) (Fig. 4.2). Like EGL-30(Gαq) mutants, EGL-8(PLCβ) mutants were first identified in a screen for egg-laying (egl) mutants that retained more eggs within adult animals than wild-type animals (Trent et al., 1983). EGL-8(PLCβ) mutant animals also moved slowly and were resistant to paralysis when exposed to aldicarb (Lackner et al., 1999). PLCβ enzymes are activated by members of the Gq/G11 trimeric G-protein α-family (Lee et al., 1992; Taylor et al., 1991) and both EGL-8(PLCβ) mutations and partial loss-of-function mutations in EGL-30(Gαq) caused similar phenotypes. However, null mutations in EGL-30(Gαq) were lethal whereas null mutations in EGL-8(PLCβ) were viable (Brundage et al., 1996; Lackner et al., 1999; Miller et al., 1999). Expression of the EGL-8(PLCβ) within the cholinergic motorneurons rescued the slow moving and aldicarb resistant phenotype of EGL-8(PLCβ) mutants. These results demonstrated that EGL-8(PLCβ) expression within the cholinergic motorneurons is necessary and sufficient for normal levels of ACh release. Exposure to phorbol ester, which mimic DAG and activate DAG signaling pathways, reversed the slow locomotion and decreased levels of ACh release of EGL-30(Gαq) and EGL-8(PLCβ) mutations (Lackner et al., 1999; Miller et al., 1999). Thus, increases in DAG signaling were epistatic to EGL-30(Gαq) and EGL-8(PLCβ) mutations consistent with a role for EGL-30(Gαq) and EGL-8(PLCβ) in stimulating the production of DAG. Phorbol ester also rescued the larval arrest phenotype of EGL-30(Gαq) null mutants suggesting that EGL-30(Gαq) was required for DAG production and that some level of DAG was required for viability (Reynolds et al., 2005).

4. UNC-73(RhoGEF) is a downstream effector of EGL-30(Gq)

Mutations in EGL-8(PLCβ) only partially suppressed the increased rate of ACh release caused by cholinergic motorneuron expression of constitutively active EGL-30(Gαq) (Lackner et al., 1999). This result suggested that EGL-30(Gαq) had at least two downstream effectors of which EGL-8(PLCβ) was one. Screens to suppress the slow growth and increased locomotion of a constitutively active EGL-30(Gαq) expressed from its own promoter (expression throughout the C. elegans nervous system and in pharyngeal muscle; Lackner et al., 1999) revealed that UNC-73(Trio) RhoGEF (an activator of small RhoGTPases) was a downstream effector of EGL-30(Gαq) (Williams et al., 2007). UNC-73(Trio) possesses two RhoGEF domains. RhoGEF1 activates Rac class RhoGTPases (henceforth referred to as the RacGEF domain) whereas RhoGEF2 activates Rho class RhoGTPases (henceforth referred to as the RhoGEF domain). Mutations in the RacGEF domain resulted in defects in neuronal pathfinding, cell migration, and engulfment of apoptotic cell corpses (deBakker et al. 2004; Siddiqui and Culotti, 1991; Steven et al., 1998). Mutations in the UNC-73 RhoGEF domain did not cause developmental defects but did cause a strong reduction in both locomotion and egg laying (Steven et al., 2005; Williams et al., 2007). Double EGL-8(PLCβ);UNC-73(RhoGEF) mutants strongly resembled EGL-30(Gαq) null mutants as both have larval lethal phenotypes. The larval lethality of double EGL-8(PLCβ);UNC-73(RhoGEF) mutants was rescued by phorbol ester the same as for EGL-30(Gαq) null mutants (Williams et al., 2007) suggesting that all three genes act upstream of DAG. UNC-73(RhoGEF) bound to EGL-30(Gαq) and coexpression of constitutively active EGL-30(Gαq) with UNC-73(RhoGEF) increased the activation of RhoA as measured by increased binding of RhoA to its downstream effector Rhotekin (Williams et al., 2007). Coexpression of a dominant negative EGL-30(Gαq) with UNC-73(RhoGEF) did not increase the activation of RhoA. These results suggested that EGL-30 (Gαq) activated RhoA GTPases via UNC-73(RhoGEF).

5. RHO-1(RhoA)

C. elegans has a single RhoA ortholog RHO-1(RhoA) (Chen and Lim, 1994). Expression of constitutively active RHO-1(RhoA) from either a heat-shock or cholinergic-specific promoter strongly increased both locomotion and ACh release (McMullan et al., 2006). Inhibition of endogenous RHO-1(RhoA) using C3 transferase (a specific inhibitor of RhoA GTPases) from either a heat-shock or a cholinergic-specific promoter strongly decreased both locomotion and ACh release. Neither an increase nor a decrease in RHO-1(RhoA) signaling altered neuronal morphology or numbers of motorneuron synapses. Hence, RHO-1(RhoA) acted within the cholinergic motorneurons to increase

ACh release from preexisting motorneuron synapses. Heat-shock expression of either wild-type UNC-73(RhoGEF) (Williams et al., 2007) or constitutively active RHO-1(G14V) (R. McMullan and S. Nurrish, unpublished data) in UNC-73(RhoGEF) mutant adults rescued the reduced locomotion defect of UNC-73(RhoGEF) mutants. Thus, the locomotion defects of UNC-73(Rho-GEF) mutants were not developmental and UNC-73(RhoGEF) acted to control locomotion via RHO-1(RhoA). Surprisingly, UNC-73(RhoGEF) mutants were only slightly resistant to aldicarb unlike other mutants that have slow locomotion such as EGL-30(Gαq) and EGL-8(PLCβ) although it is possible that increased muscle response to ACh in UNC-73(RhoGEF) mutants obscured changes in ACh release by the motorneurons (Steven et al., 2005; Williams et al., 2007). Rescue experiments indicated that expression of UNC-73(Rho-GEF) in the cholinergic motorneurons was not sufficient to rescue the slow locomotion of UNC-73(RhoGEF) mutants. In contrast, cholinergic expression of EGL-8(PLCβ) cDNA using transgenes was sufficient to rescue the slow locomotion of EGL-8(PLCβ) mutants (Lackner et al., 1999). In addition, cholinergic expression of constitutively active EGL-30(Gαq) or constitutively active RHO-1(RhoA) in the cholinergic motorneurons was sufficient to cause increased locomotion and hypersensitivity to aldicarb. These results leave open the possibility that EGL-30(Gαq) and RHO-1(RhoA) are required in cholinergic motorneurons plus other cells for wild-type rates of locomotion. In the case of EGL-8(PLCβ) it appears that this part of the EGL-30(Gαq) pathway is only required in cholinergic motorneurons and is therefore required in fewer cells that the UNC-73(RhoA) branch of the EGL-30(Gαq) pathway. This implies that in some cells EGL-30(Gαq) signals via UNC-73(RhoGEF) in an EGL-8(PLCβ)-independent manner.

6. DGK-1

The major negative regulators of DAG in all cells are the DAG kinases (DGKs). There are five classes of DGK, C. elegans has one member of each class, whereas mammals have either one (classes 3 and 5), two (class 4), or three (classes 1 and 2) (Goto et al., 2008; Merida et al., 2008; Topham, 2006). DGKs phosphorylate DAG to produce phosphatidic acid (PA), which is a second messenger in its own right (Roth, 2008). In mammals activated RhoA binds to the class 5 DAG kinase (DGKΘ) and inhibited its ability to phosphorylate DAG (Houssa et al., 1999) and this interaction is conserved in C. elegans (McMullan et al., 2006). Mutations in the C. elegans DGKΘ ortholog, DGK-1, resulted in increased locomotion, increased ACh release, and insensitivity to the neuromodulator serotonin (Nurrish et al., 1999). Mutations in DGK-1 are predicted to cause two defects, an increase in DAG and a decrease in PA. Phorbol ester mimics the increase in

DAG but it cannot be converted to PA by DGKs. As addition of phorbol ester closely mimicked the defects in DGK-1 mutants, increased locomotion and ACh release, it was the increased level of DAG that was responsible for the effects of the DGK-1 mutation.

7. EGL-30(Gαq)—A double whammy on DAG levels

Taken together these results demonstrate that EGL-30(Gαq) activates at least two separate pathways that control DAG levels (Fig. 4.2). In the first pathway EGL-30(Gαq) activates EGL-8(PLCβ) to generate DAG. In the second pathway EGL-30(Gαq) activates UNC-73(RhoGEF) and RHO-1(RhoA) leading to the inhibition of DGK-1. Inhibition of DGK-1 prevents the destruction of DAG causing DAG levels to rise. Thus, EGL-30(Gαq) controls both the production and destruction of DAG. Results suggest that the UNC-73(RhoGEF) pathway is required in more cells than the EGL-8(PLCβ) pathway. Possibly the UNC-73 (RhoGEF) pathway acts outside the cholinergic motorneurons to prevent the destruction of DAG produced by a EGL-30(Gαq)-regulated PLC other than EGL-8(PLCβ). However, thus far no other mutations in a PLC have been shown to cause locomotion defects. Alternatively, the UNC-73(RhoGEF) and RHO-1 (RhoA) branch of the EGL-30(Gαq) pathway could signal independently of DAG in some cells.

B. The GPA-12(Gα12) pathway

1. GPA-12(Gα12) and RHGF-1(RGS-RhoGEF)

The C. elegans ortholog of the Gα12 trimeric G-protein α-subunit (GPA-12) also regulates RHO-1(RhoA) to control DAG and thus levels of ACh release (Hiley et al., 2006) (Fig. 4.2). Expression of constitutively active GPA-12 (Gα12) from either heat-shock or cholinergic-specific promoters increased levels of ACh release and increased the accumulation of DAG within cholinergic motorneurons. In mammals Gα12 activate RhoA via RhoGEFs containing the RGS domain. Mutations in the single C. elegans RGS containing RhoGEF, RHGF-1(RGS-RhoGEF), blocked the ability of a constitutively active GPA-12(Gα12) to both stimulate ACh release and cause increases in DAG (Hiley et al., 2006). Expression of RHGF-1(RGS-RhoGEF) in the cholinergic cells of RHGF-1(RGS-RhoGEF) mutants restored the ability of constitutively active GPA-12(Gα12) to increase levels of ACh release. Inhibition of endogenous RHO-1(RhoA) also blocked increased ACh release caused by constitutively active GPA-12(Gα12), suggesting that GPA-12(Gα12) acted within the cholinergic motorneurons via RHGF-1(RGS-RhoGEF) to activate RHO-1(RhoA).

As with EGL-30(Gαq) signaling the activation of RHO-1 inhibited the destruction of DAG by DGK-1 leading to an increase in DAG levels. However, GPA-12 (Gα12) only altered the destruction of DAG unlike EGL-30(Gαq), which altered both production and destruction of DAG. Interestingly the GPA-12 (Gα12) pathway stimulated ACh release by only a subset of the known DAG signaling pathways (see later).

2. The GPA-12(Gα12) pathway is not active in laboratory conditions

Loss-of-function mutations in either GPA-12(Gα12) or RHGF-1(RGS-Rho-GEF) did not reduce levels of ACh release (Hiley et al., 2006), indeed currently there are no phenotypes associated with GPA-12(Gα12) or RHGF-1(RGS-RhoGEF) loss-of-function mutants. This suggests that under laboratory conditions the GPA-12(Gα12)/RHGF-1(RGS-RhoGEF) pathway is not active. It remains unknown under what conditions GPA-12(Gα12) could be activated to control levels of ACh release, possibly GPA-12 is involved in neuronal plasticity, for example learning and memory. It is interesting to note that constitutively active GPA-12(Gα12) also resulted in a temporary larval growth arrest due to inhibition of feeding (van der Linden et al., 2003). This growth arrest phenotype was suppressed by mutations in the protein kinase C (PKC)-like kinase TPA-1, which is regulated by DAG (Tabuse et al., 1989). However, mutations in TPA-1 did not suppress the ability of constitutively active GPA-12(Gα12) to increase ACh release in adults. Nor did mutations in RHGF-1 (RGS-RhoGEF) suppress the larval arrest phenotype caused by constitutively active GPA-12(Gα12) (Hiley et al., 2006). Thus, GPA-12(Gα12) appears to signal via DAG by different pathways in larval stages and in adults.

C. The GOA-1(Gαo) pathway

1. Serotonin inhibits ACh release

C. elegans exposed to exogenous serotonin moved more slowly than nonexposed animals (Choy and Thomas, 1999; Ranganathan et al., 2000; Sawin et al., 2000; Schafer et al., 1996; Segalat et al., 1995). Addition of exogenous serotonin decreased ACh release whereas decreased serotonin signaling, either due to addition of methiothepin (an antagonist of G-protein-coupled serotonin receptors) or a mutation in a gene required for serotonin release (cat-1 the vesicular monoamine transporter), increased ACh release (Nurrish et al., 1999). Thus, serotonin acts to decrease both locomotion and levels of ACh release.

2. GOA-1(Gαo)

Addition of exogenous serotonin to C. elegans caused resistance to aldicarb but did not alter muscle response to the nicotinic agonist levamisole (Nurrish et al., 1999). This data suggested that serotonin acts presynaptically at the NMJ to decrease levels of ACh release from motorneurons. A genetic screen for animals defective for serotonin signaling identified a number of mutants that included DGK-1 (discussed earlier) and GOA-1(Gαo) (Nurrish et al., 1999; Segalat et al., 1995) (Fig. 4.2). Mutations in GOA-1(Gαo) were also identified in a reverse genetic approach to study heterotrimeric G-protein signaling (Mendel et al., 1995). GOA-1(Gαo) represents the only C. elegans ortholog of both the Go and Gi class of mammalian Gα-subunits, although it is most homologous to the mammalian Goα-subunit and we will henceforth describe it as GOA-1(Gαo). Both DGK-1 and GOA-1(Gαo) mutants moved faster than wild type on food, had increased bodybends and had higher rates of ACh release than wild-type animals (Mendel et al., 1995; Nurrish et al., 1999; Segalat et al., 1995). Mutations in GOA-1(Gαo) also caused the accumulation of DAG in motorneurons at sites of neurotransmitter release (Nurrish et al., 1999). Expression of a constitutively active GOA-1(Gαo) from a heat-shock promoter gave the opposite effect: the transgenic animals moved slowly with shallow bodybends and had a lower rate of ACh release than wild-type animals. Screens for suppressors of constitutively active GOA-1(Gαo) identified mutations in DGK-1 and EAT-16(GαqGAP) suggesting that DGK-1 acted downstream of or in parallel with GOA-1(Gαo) (Hajdu-Cronin et al., 1999; Miller et al., 1999; Nurrish et al., 1999). However, attempts to coimmunoprecipitate GOA-1(Gαo) and DGK-1 proteins have been unsuccessful and DGK-1 activity did not change in animals with too much or too little GOA-1(Gαo) activity (S. Nurrish, unpublished data). Mutations in DGK-1 but not GOA-1(Gαo) suppressed the decreased locomotion caused by expression of constitutively active EGL-30(Gαq). Thus, GOA-1(Gαo) and DGK-1 appear to act in different parts of the various pathways that control motorneuron activity. In the motorneurons GOA-1(Gαo) is negatively regulated by the EGL-10(GαoGAP) and mutations in EGL-10(GαoGAP) mimic constitutively active GOA-1(Gαo) (Koelle and Horvitz, 1996). Activation of GOA-1(Gαo) increased expression of genes required for serotonin synthesis (Tanis et al., 2008). However, this does not explain why GOA-1(Gαo) mutants were strongly resistant to aldicarb, resistant to addition of exogenous serotonin and accumulated DAG at sites of neurotransmitter release in cholinergic motorneurons. How GOA-1(Gαo) inhibits DAG accumulation and thus ACh release remains to be explained. Outside the nervous system GOA-1(Gαo) acts redundantly with GPA-16 to control spindle orientation during early cell divisions (Gotta and Ahringer, 2001). In these early cell division RIC-8 also acts to control spindle orientation by acting as a GαoGEF for GOA-1(Gαo) (Afshar et al., 2004;

Couwenbergs et al., 2004; Hess et al., 2004). Thus, mutations in RIC-8 and GOA-1(Gαo) have the same phenotypes in early cell division (Miller and Rand, 2000) and opposite phenotypes in the adult nervous system. How RIC-8 switches from a GOA-1(Gαo) activator to a EGL-30(Gαq) activator during development is still unclear.

D. The DAG effector pathways

1. DAG stimulates ACh release via at least two effectors

Phorbol ester, which mimics accumulation of DAG in membranes, stimulates neurotransmitter release in both mammals and C. elegans (Malenka et al., 1986; Nurrish et al., 1999). Both DAG and phorbol ester signal by recruitment of proteins to membranes via the C1 domain present in DAG-regulated proteins. PKC is the best known example of a protein that is activated by binding to DAG or phorbol ester. Initially, it was believed that phorbol ester stimulated neurotransmitter release via PKC (Majewski and Iannazzo, 1998). However, experiments in C. elegans demonstrated that the DAG-binding protein UNC-13 was required for phorbol ester activation of neurotransmitter release (Lackner et al., 1999) and later the mammalian homologs, MUNC13-1 and MUNC13-2, were also shown to be necessary for phorbol ester activation of neurotransmitter release in mice (Rhee et al., 2002). It was argued that PKC is not required for phorbol ester activation of neurotransmitter release but recent papers suggest that both UNC-13 and PKC activation are required for the full activation of neurotransmitter release by phorbol ester (Sieburth et al., 2006; Wierda et al., 2007). In C. elegans UNC-13 and PKC-1 were activated separately of each other by phorbol ester, a mutation in either protein only partially blocked stimulation of neurotransmitter release by phorbol ester (Sieburth et al., 2006) (Fig. 4.2). In mice Wierda et al. (2007) reported a mutation in either MUNC13 or PKC completely blocked activation of neurotransmitter release by phorbol ester. However, Lou et al. (2008) reported that in mouse neurons PKC and UNC13 acted in parallel to increase neurotransmitter release in response to phorbol ester, the same as in C. elegans. Interestingly, activation of GPA-12(Gα12) stimulated ACh release via DAG signaling through UNC-13 and not PKC-1 (Hiley et al., 2006) suggesting that increases in DAG do not necessarily activate all DAG effectors. Perhaps there are separate pools of DAG that activate different DAG effectors. Alternatively, GPA-12(Gα12) may trigger a small increase in DAG and UNC-13 is more sensitive than PKC-1 to changes in DAG levels. It would be interesting to test whether any upstream activators trigger signaling via DAG-mediated activation of PKC-1 and not UNC-13.

2. PKC-1

In mice neurons, PKC phosphorylated MUNC18 to stimulate neurotransmitter release in response to phorbol ester (Wierda *et al.*, 2007). C. *elegans* UNC-18 can be phosphorylated by rat derived PKC *in vitro* (Sassa *et al.*, 1996) although the effect of these phosphorylations on UNC-18 function have not been determined. In C. *elegans* activation of neurotransmitter release via PKC-1 was reported to be independent of UNC-18 phosphorylation (Sieburth *et al.*, 2007), which is surprising given the close similarity of DAG signaling in both C. *elegans* and mammalian neurons. In C. *elegans* phorbol ester stimulates release from neurosecretory cells (Zhou *et al.*, 2007) and mutations in PKC-1 decreased the release of GFP-tagged NLP-21 neuropeptide from the cholinergic motorneurons (Sieburth *et al.*, 2007) suggesting that DAG acts via PKC-1 (and possibly UNC-13 as well) to stimulate release of neuropeptides from dense-core vesicles (DCVs). Mutations in PKC-1 also caused resistance to aldicarb suggesting a decrease in levels of ACh signaling. Other results suggested that release of neuropeptides or other DCV contents also altered levels of ACh signaling. For example, mutations in genes required for DCV release [UNC-31(CAPS)] or processing of neuropeptides [EGL-3(Prohormone convertase2) or EGL-21(Carboxypeptidase E)] caused resistance to aldicarb (Jacob and Kaplan, 2003; Kass *et al.*, 2001; Miller *et al.*, 1996). However, in the case of EGL-3(Prohormone convertase2) or EGL-21(Carboxypeptidase E) mutants, rescue of neuropeptide release from only the cholinergic motorneurons and interneurons was not sufficient to restore normal ACh signaling at the NMJ (Jacob and Kaplan, 2003). Despite the aldicarb resistance of PKC-1 mutants, electrophysiological recordings of C. *elegans* body-wall muscle in PKC-1 mutants revealed no defects in either synaptic vesicle release or muscle response to ACh (Sieburth *et al.*, 2007). How is it then that PKC-1 mutants were resistant to aldicarb, caused behavioral defects and altered neuropeptide release from cholinergic motorneurons without altering ACh signaling at the body-wall muscles? It is possible that while preparing worms for electrophysiology neuropeptides were lost and their effect was absent. These results suggest that the electrophysiological preparation at the C. *elegans* NMJ requires further investigation as to its reliability.

3. UNC-13

UNC-13 is essential for neurotransmitter release in C. *elegans*, *Drosophila*, and mammals (Aravamudan *et al.*, 1999; Augustin *et al.*, 1999; Richmond *et al.*, 1999). Three MUNC13 genes are present in mammals but there is only one UNC-13 gene in C. *elegans* although it is alternatively spliced to generate different isoforms (Kohn *et al.*, 2000). The longer UNC-13 protein in C. *elegans* binds to the synaptic release site localized UNC-10(RIM) (Betz *et al.*, 2001;

Koushika *et al.*, 2001; Schoch *et al.*, 2002). Shorter forms of UNC-13 lack the UNC-10(RIM)-binding site. However, all the isoforms of UNC-13 contain the C1 domain, which binds DAG. It was suggested that some of the long forms of UNC-13 are always present at release sites via binding to UNC-10(RIM) and this UNC-13 maintains basal levels of neurotransmitter release. Upon production of DAG at release sites additional long and short forms of UNC-13 are recruited to release sites to increase levels of neurotransmitter release (Lackner *et al.*, 1999; Nurrish *et al.*, 1999). However, in mammals Brose has argued that sufficient levels of MUNC13 for stimulated neurotransmitter release are always present at release sites and that binding of the MUNC13 C1 domain to DAG causes a conformational change that either increases the ability of UNC-13 to stimulate release, or results in an additional function of UNC-13 that stimulates some other aspect of release, or a combination of these two possibilities (Basu *et al.*, 2007). UNC-13 is believed to act in part by interacting with the SNARE protein UNC-64(syntaxin) (Betz *et al.*, 1997). UNC-13 binding to UNC-64 (syntaxin) was essential for UNC-13′s ability to promote neurotransmitter release in C. *elegans* and for chromaffin granule priming in mammals (Madison *et al.*, 2005; Stevens *et al.*, 2005). UNC-64(syntaxin) can exist in two folded states referred to as closed and open (Dulubova *et al.*, 1999). Only open UNC-64 (syntaxin) can bind to other SNARE proteins (synaptobrevin and SNAP25) and thus stimulate neurotransmitter release. The role of UNC-13 has been proposed to be to bind UNC-64(syntaxin) present at release sites and increase the ratio of open UNC-64(syntaxin) and thus increase levels of neurotransmitter release. Open-form mutations that biased UNC-64(syntaxin) towards the open state were reported to restore neurotransmitter release to UNC-13 mutants in C. *elegans* (Richmond *et al.*, 2001). Other reports have suggested that open-form UNC-64(syntaxin) restored priming defects in UNC-13 mutants but did not restore neurotransmitter release (McEwen *et al.*, 2006). It is possible that UNC-13 has multiple essential functions during neurotransmitter release and activating UNC-64(syntaxin) is only one. Regions of UNC-13 outside of the UNC-64(syntaxin) interaction domain blocked binding of full-length UNC-13 to UNC-64(syntaxin) (Basu *et al.*, 2005; Madison *et al.*, 2005). It seems likely that other as yet unknown factors or posttranslational modifications of UNC-13 are required for binding of full-length UNC-13 to UNC-64(syntaxin). Thus, it appears there is much more to discover about how UNC-13 is regulated and what its essential roles in neurotransmitter release are.

4. DAG-independent pathways also stimulate ACh release

Three pieces of data demonstrated that RHO-1(RhoA) could increase ACh release and rates of locomotion independent of DGK-1 (McMullan *et al.*, 2006). Firstly, a mutation in RHO-1(RhoA), F25N, almost completely abolished the ability of RHO-1(RhoA) to bind DGK-1. However, the F25N RHO-1 (RhoA) mutant was still able to increase locomotion and ACh release although with reduced efficiency. Secondly, inhibition of RHO-1(RhoA) by C3 transferase reduced locomotion and reduced ACh release. If RHO-1(RhoA) had mediated all of its effects on ACh signaling by inhibition of DGK-1 then inhibition of RHO-1(RhoA) should have had no effect in a DGK-1 mutant where DGK-1 activity would be already fully inhibited. However, in a DGK-1 mutant C3 transferase still partially inhibited rates of locomotion and ACh release. Finally, mutations that blocked the DAG responsiveness of the DAG effectors UNC-13 and PKC-1 did not completely block the ability of constitutively active RHO-1(RhoA) to increase levels of ACh release (R. McMullan and S. Nurrish, unpublished data). Thus, there are at least two pathways downstream of RHO-1(RhoA) that act in the cholinergic motorneurons to increase locomotion and ACh release. One attractive candidate is the pathway controlling actin given the known role of RhoGTPases in rearranging the cytoskeleton. Changes in actin have been associated with changes in neuronal activity in mammalian neurons (Dillon and Goda, 2005; Hall, 1998). However, as yet there is no evidence to point to a role for actin in control of C. *elegans* neurotransmission.

E. The GSA-1(Gαs) pathway

1. GSA-1(Gαs) and adenylate cyclase

A screen to restore normal locomotion to slow moving RIC-8 mutants identified gain-of-function mutations in GSA-1(Gαs) and its downstream effector ACY-1 (adenylate cyclase) (Schade *et al.*, 2005) (Fig. 4.2). GSA-1(Gαs) gain-of-function mutations moved at faster rates than wild-type animals and this was completely suppressed by loss-of-function mutations in ACY-1(adenylate cyclase). Thus, ACY-1(adenylate cyclase) acted downstream of GSA-1(Gαs) and GSA-1(Gαs) could not increase signaling at the NMJ in the absence of ACY-1(adenylate cyclase). This is consistent with the known ability of mammalian Gαs to stimulate adenylate cyclases to produce cAMP (Schade *et al.*, 2005). Heat-shock expression of a GSA-1(Gαs) gain-of-function mutation in adults also increased locomotion indicating that the effects of GSA-1(Gαs) on locomotion were not due to developmental changes. Animals carrying either a GSA-1 (Gαs) gain-of-function mutation or a ACY-1(adenylate cyclase) gain-of-function mutation were hypersensitive to aldicarb suggesting increased levels of ACh release from the motorneurons. In addition both mutations also caused animals to be resistant to paralysis by levamisole, suggesting that the muscles were less sensitive to ACh, which could cause the size of an increase in ACh to be underestimated by the aldicarb assay. Changes in the muscle could possibly be due to homeostasis where the muscle adapted to excess levels of ACh release by the cholinergic motorneurons or it could reflect signaling by GSA-1(Gαs) via ACY-1(adenylate cyclase) within the muscles themselves. Both GSA-1(Gαs) and ACY-1(adenylate cyclase) were expressed in both neurons and body-wall muscles (Berger et al., 1998). Expression of ACY-1(adenylate cyclase) gain-of-function mutants in the body-wall muscles, but not the neurons, caused animals to become hypersensitive to levamisole, the opposite result to that observed in animals with a genomic gain-of-function mutation in ACY-1(adenylate cyclase) (Schade et al., 2005). These results may have been due to an overexpression artifact or perhaps different levels of signaling via the GSA-1 (Gαs) pathway can have different effects in muscle. Transgenic overexpression of an ACY-1(adenylate cyclase) gain-of-function mutation in muscles and neurons at the same time resulted in animals with a normal response to levamisole, that is, their muscles had a normal response to ACh. These results suggested that the strength of signaling through the GSA-1(Gαs) pathway is important and that the GSA-1(Gαs) pathways in both neurons and muscles are somehow communicating with each other. Transgenic expression of the ACY-1(adenylate cyclase) gain-of-function mutation in only the nervous system or only the body-wall muscles caused increased rates of locomotion but neither caused any change in sensitivity to aldicarb. Only transgenic expression of the ACY-1(adenylate cyclase) gain-of-function mutation from both the neuronal and body-wall-specific promoters in the same animal was sufficient to cause hypersensitivity to aldicarb.

2. PDE-4 cAMP phosphodiesterase

The model for GSA-1(Gαs) regulation of NMJ activity is that GSA-1(Gαs) activates of ACY-1(adenylate cyclase) leading to the production of cyclic adenosine monophosphate (cAMP), which goes onto increase activity at the NMJ. This model was strongly supported by the finding that mutations in the homolog of the Dunce cAMP phosphodiesterase PDE-4(Dunce), which removes cAMP, caused increased rates of locomotion although the response to aldicarb was the same as for wild-type animals (Charlie et al., 2006b). PDE-4(Dunce) was predicted to decrease levels of cAMP and overexpression of PDE-4(Dunce) in neurons did decrease rates of locomotion. Thus, the level of cAMP within neurons appeared to correlate with locomotion rate, the more cAMP the faster the locomotion. However, increasing cAMP levels by exposure to membrane permeable cAMP analogs caused animals to reduce their locomotion rate in contrast to changes in GSA-1(Gαs) signaling that increase cAMP (Schade et al., 2005). This could have been because addition of cAMP analogs increased levels of cAMP above that possible by activation of the GSA-1(Gαs) pathway, or that raising cAMP in the absence of activated GSA-1(Gαs) had different down-stream effects, or that spatially restricted increases in cAMP within neurons is important, or a combination of these possibilities. A mutation in PDE-4(Dunce) was still able to increase rates of locomotion in null mutants of ACY-1(adenylate cyclase) (Charlie et al., 2006b). As ACY-1(adenylate cyclase) mutations completely block the ability of GSA-1(Gαs) to increase locomotion these results suggest that PDE-4(Dunce) degraded cAMP that was generated independently of the GSA-1(Gαs) signaling pathway, the signaling pathway responsible for making this cAMP remains unknown.

3. Protein kinase A

Another suppressor of RIC-8 was a loss-of-function mutation in the KIN-2 (PKAr) gene (Schade et al., 2005). KIN-2(PKAr) encodes the regulatory subunit of protein kinase A (PKA), which binds to and inactivates KIN-1(PKAc) the PKA catalytic subunit of PKA. Activation of adenylate cyclase produces cAMP, which then binds to KIN-2(PKAr). Binding of cAMP to KIN-2(PKAr) causes KIN-1(PKAc) and KIN-2(PKAr) to disassociate allowing KIN-1(PKAc) to become active and phosphorylate substrate proteins (Schade et al., 2005). Loss-of-function mutations in KIN-2(PKAr) caused very similar phenotypes as gain-of-function mutations in either GSA-1(Gαs) or ACY-1(adenylate cyclase) suggesting that the GSA-1(Gαs) pathway acts to disassociate KIN-2(PKAr) from KIN-1(PKAc). The KIN-1(PKAc) gene produces numerous proteins with alternate N and C termini (Gross et al., 1990; Tabish et al., 1999). RNAi knockdown of one of the six alternate N terminal splice forms led to defects in locomotion whereas knockdown of a different splice form had no effect (Murray et al., 2008). These results are consistent with a GSA-1(Gαs) signaling pathway that modulates activity at the NMJ by activating ACY-1(adenylate cyclase) to increase levels of cAMP leading to activation of KIN-1(PKAc). Thus, synaptic activity in both C. elegans and mammalian neurons is regulated by activation of PKA (Evans and Morgan, 2003; Nguyen and Woo, 2003). The relevant sub-strates of KIN-1(PKAc) at the C. elegans NMJ are unknown. In mammals PKA often signals to the transcription factor CREB. However, C. elegans CREB, CRH-1(CREB), is not expressed in either the motorneurons or the body-wall muscles and mutants moved relatively normally (Kimura et al., 2002). Another target of PKA in mammalian neurons are synapsins (Hilfiker et al., 1999) and RNAi knockdown of SNN-1(synapsin) did cause a decrease in ACh release (Sieburth et al., 2005). It will be interesting to test whether SNN-1(synapsin) mutants can block the increased locomotion and ACh release caused by increased GSA-1(Gαs) signaling.

4. UNC-31(CAPS)

UNC-31(CAPS) is required for release of DCV contents and mutations in UNC-31(CAPS) alter a number of *C. elegans* behaviors (Avery *et al.*, 1993; Brenner, 1974). The role of UNC-31(CAPS) in the release of DCV contents is unclear; proposed roles have included filling of the DCVs (Liu *et al.*, 2008; Speidel *et al.*, 2005), uptake of monoamines (Brunk *et al.*, 2009), DCV cargo stability (Speidel *et al.*, 2008), and DCV exocytosis (Ann *et al.*, 1997; Elhamdani *et al.*, 1999; Hammarlund *et al.*, 2008). A role for UNC-31(CAPS) in synaptic vesicle release has been controversial with evidence both for (Jockusch *et al.*, 2007) and against (Gracheva *et al.*, 2007b; Hammarlund *et al.*, 2008; Speese *et al.*, 2007; Tandon *et al.*, 1998). In the presence of food, mutations in either UNC-31(CAPS) or ACY-1(adenylate cyclase) were almost paralyzed and double mutants had the same phenotype as single mutants (Charlie *et al.*, 2006a). Animals mutant for UNC-31(CAPS) or ACY-1(adenylate cyclase) also had a normal or only a slight resistance to aldicarb suggesting normal levels of ACh release. This was surprising as UNC-31(CAPS) mutations were recovered in screens for resistance to aldicarb (Miller *et al.*, 1996) and UNC-31(CAPS) mutants were almost paralyzed in the presence of food. However, in the absence of food UNC-31(CAPS) mutants moved normally suggesting that UNC-31 (CAPS) mutants have the ability to move but on food they lack the motivation to move (Avery *et al.*, 1993). UNC-31(CAPS) was required in the cholinergic motorneurons for normal locomotion suggesting that DCV release from the cholinergic motorneurons was required for normal signaling at the NMJ (Charlie *et al.*, 2006a). The increased locomotion and ACh release of GSA-1 (Gαs) gain-of-function mutations was not blocked by mutations in UNC-31 (CAPS). Activation of the GSA-1(Gαs) pathway only in the cholinergic motorneurons, but not the body-wall muscles, restored the locomotion of UNC-31(CAPS) mutant animals almost to wild type. Given that UNC-31 (CAPS) was thought to be essential for DCV release it was suggested that DCVs in the motorneurons were releasing contents that were triggering activation of the GSA-1(Gαs) pathway also in the motorneurons (autocrine signaling). However, recently it was shown that activation of KIN-1(PKAc) could bypass the requirement for UNC-31 in DCV release (Zhou *et al.*, 2007) although the target of KIN-1(PKAc) in DCV release is unknown. Thus, the GSA-1(Gαs) pathway and UNC-31(CAPS) pathway probably converge to control DCV release. It remains unclear what NMJ neuroregulators are being released by the DCVs in response to GSA-1(Gαs) signaling and whether these released neuroregulators target receptors on motorneurons, body-wall muscles, or both.

5. GSA-1(Gαs) and the EGL-30(Gαq) pathway

Mutations in UNC-31(CAPS) partially suppressed the increased locomotion but not the increased ACh release observed in GOA-1(Gαo) loss-of-function mutations and in EGL-30(Gαq) gain-of-function mutations (Charlie *et al.*, 2006a). Again this suggests that rates of locomotion and rates of ACh release do not always correlate. Activation of the GSA-1(Gαs) pathway only slightly increased locomotion rate in strong EGL-30(Gαq) mutants. These results suggested that the GSA-1(Gαs) pathway cannot increase activity at the NMJ in the absence of the EGL-30(Gαq) pathway. Combining EGL-30(Gαq) and UNC-31(CAPS) loss-of-function mutations reduced the rate of locomotion below that of either single mutant suggesting that the EGL-30(Gαq) and GSA-1(Gαs) pathways acted in parallel in the motorneurons (Charlie *et al.*, 2006a). Activation of the EGL-30(Gαq) pathway (using EGL-30(Gαq) gain-of-function mutants or by exposure to phorbol ester) in GSA-1(Gαs) mutants increased locomotion rates but not as much as in animals where the GSA-1(Gαs) signaling pathway was intact (Reynolds *et al.*, 2005). Thus, mutations in the GSA-1(Gαs) pathway partially blocked the effects of phorbol ester implying that GSA-1(Gαs) must, at least in part, act at a step independent of DAG activation of synaptic vesicle and DCV release. If the GSA-1(Gαs) pathway does stimulate DCV release then it acts at a different step to PKC-1, which also acts on DCV release, as PKC-1 mutants are fully rescued by phorbol ester (Sieburth *et al.*, 2007).

6. RIC-8

The relationship between RIC-8 and GSA-1(Gαs) is unclear as RIC-8 can act as a GαGEF able to activate GSA-1(Gαs), EGL-30(Gαq), and GOA-1(Gαo) (Klattenhoff *et al.*, 2003; Romo *et al.*, 2008; Tall *et al.*, 2003). In many respects RIC-8 null mutants most closely resembled EGL-30(Gαq) mutants, both were almost completely paralyzed and paralysis was not significantly suppressed by activation of the GSA-1(Gαs) pathway (Reynolds *et al.*, 2005). However, addition of phorbol ester to EGL-30(Gαq) mutants completely rescued locomotion but only partially rescued the locomotion of RIC-8 null mutants, this is characteristic of signaling via GSA-1(Gαs). Thus, it appears that RIC-8 acts on both the EGL-30(Gαq) and GSA-1(Gαs) pathways. It is possible that RIC-8 also activates signaling by GOA-1(Gαo) but that RIC-8 effects on the EGL-30 (Gαq) and GSA-1(Gαs) pathway obscures this.

F. Other heterotrimeric G-protein subunits and regulators

1. Heterotrimeric Gβ-subunits

Heterotrimeric G-proteins consist of three subunits: α, β, and γ. There are 21 α-subunits although most are expressed only in the chemosensory neurons and not in the motorneurons (Bastiani and Mendel, 2006; Jansen et al., 1999). The only Gα-subunits known to have a role in NMJ activity are EGL-30(Gαq), GPA-12 (Gα12), GOA-1(Gαo), and GSA-1(Gαs). The C. elegans genome contains two Gβ-subunits, GPB-1 and GPB-2. GPB-1 is most similar to canonical mammalian Bβ-subunits, is widely expressed in most if not all cells, and available mutations are lethal (Zwaal et al., 1996). Overexpression of GPB-1 resulted in lethargic locomotion whereas some mosaic animals that had lost GPB-1 in a subset of cells moved with exaggerated bodybends. The most likely explanation of these results is that GPB-1 overexpression titrates out EGL-30(Gαq) subunits preventing them from activating their downstream activators (van der Linden et al., 2001). GPB-2 is most closely related to the mammalian Gβ5-subunit, it is expressed both neuronally and in muscles, and GPB-2 mutations are viable (Chase et al., 2001; Robatzek et al., 2001; van der Linden et al., 2001). Animals carrying a GPB-2 mutation had increased ACh release and were hypersensitive to starvation and arecoline (van der Linden et al., 2001; You et al., 2006). Double GPB-2;GOA-1(Gαo) mutants had a synthetic larval arrest phenotype not observed in either single mutant and this was suppressed by mutations in EGL-30 (Gαq) (van der Linden et al., 2001). This suggests that loss of GOA-1(Gαo) and GPB-2 causes a lethal hyperactivation of EGL-30(Gαq); however, it is unclear whether lethality was caused by hyperactivation of EGL-30(Gαq) in the cholinergic motorneurons or other cells. GPB-2 interacts with both EAT-16 (GαqGAP) and EGL-10(GαoGAP) as well as both Gγ-subunits (see later) (van der Linden et al., 2001) and the loss of either GPB-2 or EGL-10(GαoGAP) caused a reduction in the protein levels of the other (Chase et al., 2001). Double mutants of GPB-2;EAT-16(GαqGAP), or of GPB-2;EGL-10(GαoGAP) resembled GPB-2 single mutants in some reports (van der Linden et al., 2001) but not in others (Chase et al., 2001). These differing results in the literature make it difficult to establish a model for GPB-2 in control of activity at the NMJ. Both groups did report that a double EAT-16(GαqGAP);EGL-10(GαoGAP) mutant closely resembles the GPB-2 mutant. These results suggested that GPB-2 binds to both EAT-16(GαqGAP) and EGL-10(GαoGAP) and is required for their ability to inactive EGL-30(Gαq) and GOA-1(Gαo) respectively. It was suggested that the sole function of GOA-1(Gαo) within the motorneurons is to activate EAT-16(GαqGAP) and thus negatively regulate signaling of EGL-30

(Gαq). However, it has not been shown that the GAP activity of EAT-16 (GαqGAP) towards EGL-30(Gαq) is regulated in a GOA-1(Gαo)-dependent manner.

2. Heterotrimeric Gγ-subunits

C. *elegans* has two Gγ-subunits, GPC-1 and GPC-2 (Jansen *et al.*, 1999). GPC-1 expression is limited to sensory neurons whereas GPC-2 is expressed in all neurons and muscles and therefore GPC-2 is most likely to represent the Gγ-subunit used for all G-protein signaling at the NMJ (Jansen *et al.*, 2002). GPC-1 mutants were viable and were defective for olfactory adaptation (Hilliard *et al.*, 2005; Jansen *et al.*, 2002; Yamada *et al.*, 2009). RNAi knockdown of GPC-2 caused early embryonic lethality (Gotta and Ahringer, 2001).

3. Arrestin and GRKs

In mammals GPCR signaling is negatively regulated by arrestin and GPCR kinases (GRK). C. *elegans* has a single ARR-1(arrestin) and it is widely expressed in many if not all neurons and some muscles (Fukuto *et al.*, 2004; Palmitessa *et al.*, 2005). Mutations in ARR-1(arrestin) were viable; however, the mutants had no obvious defects in locomotion (Palmitessa *et al.*, 2005) although RNAi knockdown of ARR-1(arrestin) did cause resistance to aldicarb suggesting a decrease in ACh signaling (Sieburth *et al.*, 2005). The C. *elegans* genome encodes two GRKs. GRK-1 is most similar to human GRK5 whereas GRK-2 is most similar to human GRK3 and GRK2 (Fukuto *et al.*, 2004). GRK-2 mutations caused defects in chemotaxis to certain odorants but no defects in locomotion were reported (Fukuto *et al.*, 2004). RNAi knockdown of either GRK-1 or GRK-2 did not alter sensitivity to aldicarb although it is unclear whether protein levels of either were affected (Sieburth *et al.*, 2005). It is perhaps surprising that neither GRK-1 nor GRK-2 had any major effects on locomotion, and defects in ARR-1 (arrestin) appeared to have only subtle effects on locomotion. Possibly arrestins and GRKs mainly act to regulate G-protein signaling in chemosensory neurons and not motorneurons.

G. G-protein-coupled receptors

1. mAChRs

With the possible exception of RIC-8 the activation of heterotrimeric G-proteins requires the binding of ligand to GPCRs. Surprisingly, we know very little about the receptors that activate EGL-30(Gαq), GPA-12(Gα12),

GOA-1(Gαo), or GSA-1(Gαs) in the cholinergic motorneurons. Genetic screens for resistance to aldicarb, serotonin, and arecoline have not identified mutations in GPCRs. The most obvious example of a GPCR expected to regulate ACh release are muscarinic ACh receptors (mAChRs), which are targets for arecoline and atropine (Avery, 1993; Brundage et al., 1996; Hawasli et al., 2004; Lackner et al., 1999; Steger and Avery, 2004; You et al., 2006). There are three mAChRs in the C. elegans genome, GAR-1, GAR-2, and GAR-3 (Hwang et al., 1999; Lee et al., 1999b, 2000; Park et al., 2000; Suh et al., 2001). Only GAR-2 is known to be expressed in motorneurons in the ventral nerve chord (Lee et al., 2000). GAR-2 most closely resembled the M2/M4 class of mAChRs that are coupled to Go class of G-proteins. A lack of GAR-2 signaling in the cholinergic motorneurons caused an increase in ACh release that resembled the increase in ACh release in GOA-1(Gαo) mutants (Dittman and Kaplan, 2008). Over-expression of GAR-2 in the cholinergic motorneurons strongly decreased ACh release and this was completely reversed by mutations in GOA-1(Gαo). Thus, GAR-2 activated GOA-1(Gαo) in the cholinergic motorneurons. However, the results using arecoline and atropine suggested that at least one mAChR was coupled to EGL-30(Gαq). Overexpression of GAR-3 in the cholinergic motor-neurons did increase ACh release; however, there is no evidence that GAR-3 is expressed in the cholinergic motorneurons that innervate the body-wall muscles. Thus, there is currently no evidence for a mAChR present on the cholinergic motorneurons stimulating EGL-30(Gαq) to increase ACh release. It was possible that mAChR present on other cells caused the release of neuromodulators that modulated motorneuron activity. Nonetheless the evidence suggests that GPCRs coupled to EGL-30(Gαq) must be present on the cholinergic motorneurons, although they might not necessarily be mAChRs.

2. GPCRs coupled to GOA-1(Gαo)

Serotonin negatively regulated ACh release onto the body-wall muscles in a manner dependent on GOA-1(Gαo) (Nurrish et al., 1999). The simplest expla-nation for this data was that a serotonin GPCR in cholinergic motorneurons was coupled to GOA-1(Gαo). C. elegans possesses between three and eight serotonin GPCRs, the exact number is complicated by the existence of tyramine and octopamine receptors absent in mammals (Carre-Pierrat et al., 2006). The literature on C. elegans GPCRs is confusing. For example, the same SER-1 (5HTR) mutation has been reported to be strongly resistant to serotonin induced slowing of locomotion (Dernovici et al., 2007) or to have had no effect at all (Carre-Pierrat et al., 2006). A further complication was that the nematode-specific MOD-1 serotonin-gated chloride channel was also required for serotonin-mediated effects on locomotion, although the site of action for

MOD-1 was never established (Ranganathan *et al.*, 2000). Thus, it remains unclear which serotonin receptors were acting upstream of GOA-1(Gαo) to decrease ACh release. It is possible serotonin acts on receptors in cells other than the cholinergic motorneurons and that these cells then released a different neuromodulator that bound to receptors present on the cholinergic motorneurons. One such neuromodulator could have been FLP-1(FMRFamide), which is a homolog of the FMRFamide-related neuropeptide family. Mutations in FLP-1 (FMRFamide) resulted in hyperactive locomotion that was suppressed by EGL-30 (Gαq) mutations (Nelson *et al.*, 1998). Overexpression of FLP-1 resulted in lethargic locomotion that was suppressed by mutations in GOA-1(Gαo). The simplest explanation of these results was that biologically active peptides generated from the FLP-1 propeptide (Husson *et al.*, 2007; Rosoff *et al.*, 1993) were released and bound to receptors, possibly GPCRs coupled to GOA-1(Gαo), on the cholinergic motorneurons, to regulate activity at the NMJ. C. *elegans* neuropeptides have been shown to act directly on the pharyngeal muscle to either increase or decrease muscle activity and it seems likely that neuropeptides will also directly bind GPCRs present on bodywall muscles to alter muscle activity (Papaioannou *et al.*, 2005, 2008; Rogers *et al.*, 2001). Other GPCRs that negatively regulated ACh release were GBB-1 and GBB-2 that formed a heterodimer to mediate GABA signaling at the cholinergic motorneurons (Dittman and Kaplan, 2008). Mutations in either GBB-1 or GBB-2 increased both locomotion and ACh release. GABA is released to relax muscles opposite to the side where muscles are contracting during locomotion. It was possible that GABA release not only acted directly on muscles but also on adjacent cholinergic motorneurons to reduce ACh release and ensure muscle relaxation. Given the inhibitory effect of GBB-1 and GBB-2 on ACh release it is likely they were coupled to GOA-1(Gαo).

3. Screens for GPCRs

RNAi knockdown of 60 predicted GPCRs identified seven that result in abnormal locomotion (Keating *et al.*, 2003). For example, NPR-3 was expressed in the motorneurons and RNAi knockdown resulted in very lethargic locomotion with shallow bodybends. These results suggested that in C. *elegans* NPR-3 was coupled to EGL-30(Gαq) receptors; however, expression of NPR-3 in CHO cells suggested that NPR-3 is coupled to Gαo (Kubiak *et al.*, 2003).

4. Why haven't we found the GPCRs?

The most outstanding problem in our understanding of how cholinergic motorneurons are regulated is what are the extracellular neuromodulators and the GPCR receptors they bind to activate the EGL-30(Gαq), GPA-12(Gα12), GOA-1(Gαo), and GSA-1(Gαs) signaling pathways. *C. elegans* contains at least 113 neuropeptide genes predicted to encode over 250 possible neuropeptides after processing (Li and Kim, 2008). In addition, the *C. elegans* genome is predicted to encode over 1300 GPCRs along with about 400 GPCR pseudogenes, which is roughly 7% of all the protein encoding genes (Thomas and Robertson, 2008). Although the majority of these GPCRs are believed to function in chemosensory neurons (Bargmann, 2006), it seems likely that many GPCRs are present on motorneurons. If the function of GPCRs expressed in motorneurons overlap it will be difficult to demonstrate which GPCRs mediate neuromodulator signaling to the EGL-30(Gαq), GPA-12(Gα12), GOA-1(Gαo), and GSA-1(Gαs) signaling pathways.

H. Other genes that alter motorneuron activity

Many other genes that regulate the activity of the NMJ have not been placed into any of the EGL-30(Gαq), GPA-12(Gα12), GOA-1(Gαo), or GSA-1(Gαs) signaling pathways (Nurrish, 2002; Richmond and Broadie, 2002). Many of these genes represent essential components of all neurotransmitter release, whereas a smaller number are specifically required for ACh signaling (Brenner, 1974; Hosono and Kamiya, 1991; Miller *et al.*, 1996; Nguyen *et al.*, 1995). Many essential genes for neurotransmitter release in both *C. elegans* and mammals were first identified in *C. elegans* in screens for aldicarb resistance.

1. Exocytosis

Synaptic vesicle fusion requires the interaction of three families of SNARE (SNAP and NSF attachment receptors) proteins (Sudhof and Rothman, 2009). Mutations in one member of each family have been shown to cause defects in ACh release: UNC-64(syntaxin) (Saifee *et al.*, 1998), SNB-1(synaptobrevin) (Nonet *et al.*, 1998), and RIC-4(SNAP25) (Hwang and Lee, 2003). As mentioned earlier UNC-64(syntaxin) is positively regulated by UNC-13 (McEwen *et al.*, 2006; Richmond *et al.*, 2001) and mutations in either gene caused a severe defect in ACh release. UNC-64(syntaxin) is also the target of many other proteins that are required for normal levels of ACh release. The correct functioning of UNC-64(syntaxin) is dependent on the neuronal-specific UNC-18 protein and mutations in UNC-18 cause severe defects in locomotion and ACh release (Brenner, 1974; Gengyo-Ando *et al.*, 1993; Hosono and Kamiya, 1991; Hosono *et al.*, 1992; Miller *et al.*, 1996; Nguyen *et al.*, 1995; Sassa *et al.*, 1999). Trafficking of UNC-64(syntaxin) from the endoplasmic reticulum to synapses has been shown to be dependent on UNC-18 (McEwen and Kaplan, 2008) although this was not observed by others (Weimer *et al.*, 2003). Both UNC-64(syntaxin) and UNC-18 were required at synapses for correct docking of synaptic vesicles to the plasma membrane (Hammarlund et al., 2007; Weimer et al., 2003) and they must be bound to each other for normal levels of neurotransmitter release (Johnson et al., 2009). UNC-64(syntaxin) is negatively regulated by TOM-1(tomosyn), which competes with SNB-1 (synaptobrevin) for binding to UNC-64(syntaxin) (Dybbs et al., 2005; Fujita et al., 1998; Gracheva et al., 2006, 2007a; McEwen et al., 2006). Mutations in TOM-1(tomosyn) caused increased levels of ACh release and increased locomotion consistent with a role for TOM-1(tomosyn) as an inhibitor of synaptic vesicle release. TOM-1(tomosyn) mutants also appeared to alter DCV exocytosis although the mechanism remains unclear (Gracheva et al., 2007b). The interaction of mammalian TOM-1(tomosyn) with UNC-64(syntaxin) was positively regulated by phosphorylation of UNC-64(syntaxin) by the Rho-regulated kinase ROCK (Sakisaka et al., 2004); however, it is unclear if LET-502(ROCK) in C. elegans regulates motorneuron activity.

2. Ion channels

Evoked release of synaptic vesicles is dependent on the entry of Ca^{2+} into nerve terminals upon depolarization and changes in the activity of Ca^{2+}, K^+, and Na^+ ion channels can strongly influence neuronal activity (Hille, 2001). In mammalian neurons there is extensive regulation of ion channels by G-protein signaling pathways (Dolphin, 2006). In Drosophila phototransduction, $G\alpha q$ signaling activates Ca^{2+} influx via the TRP-type Ca^{2+} channels (Hardie, 2003). In C. elegans UNC-2 and UNC-36 both encode components of voltage-gated Ca^{2+} channels and mutations in either gene resulted in severe defects in locomotion and ACh release (Brenner, 1974; Hosono and Kamiya, 1991; Nguyen et al., 1995; Schafer and Kenyon, 1995). Partial loss-of-function mutations in UNC-64(syntaxin) were suppressed by the absence of the K^+ channel SLO-1 in neurons (Wang et al., 2001). SLO-1 mutants were hypersensitive to aldicarb, suggesting increased levels of ACh signaling at the NMJ. SLO-1 is a K^+ channel activated by Ca^{2+} via the UNC-43(CaMKII) (see later) (Liu et al., 2007). The role of SLO-1 is believed to be to terminate release of neurotransmitter after opening of voltage-gated Ca^{2+} channels. However, there is no current evidence that the EGL-30($G\alpha q$), GPA-12($G\alpha 12$), GOA-1($G\alpha o$), or GSA-1($G\alpha s$) signaling pathways exert their effects on NMJ activity via modulation of ion channels. A genetic interaction has been shown between GSA-1 ($G\alpha s$) and UNC-36 (Berger et al., 1998) in neurodegeneration although others failed to see this interaction (Korswagen et al., 1997). It seems unlikely that heterotrimeric G-proteins regulate ion channels in mammals and Drosophila neurons but not C. elegans neurons. This may reflect the difficulties in performing electrophysiology in C. elegans compared to mammalian preparations or the

Drosophila compound eye. However, it is still surprising that genetic screens have not linked the components of the known heterotrimeric G-protein signaling pathways in the motorneuron to changes in ion channel activity.

3. Docking

A group of interacting genes are also important for steps in the synaptic vesicle cycle prior to the assembly of the SNARE complex. Mutations in UNC-10(Rim) (Koushika *et al.*, 2001), AEX-3(RabGEF) (Iwasaki *et al.*, 1997), CAB-1 (NPDC1) (Iwasaki and Toyonaga, 2000), RAB-3(Rab3) (Nonet *et al.*, 1997), AEX-6(Rab27), and RBF-1(Rabphilin) (Mahoney *et al.*, 2006) caused either locomotion defects and/or resistance to aldicarb suggesting defects in signaling at the NMJ. UNC-10(Rim) is a multifunctional protein that interacts with many synaptic proteins including UNC-13(MUNC13), RAB-3(Rab3), and Ca^{2+} channels (Betz *et al.*, 2001; Coppola *et al.*, 2001; Dulubova *et al.*, 2005; Kiyonaka *et al.*, 2007; Schoch *et al.*, 2002; Wang *et al.*, 1997). Mutations in UNC-10(RIM) or RAB-3(Rab3) are believed to cause defects in the docking of synaptic vesicles to the plasma membrane. Currently, it is believed that RAB-3 (Rab3) is associated with synaptic vesicles and interaction of RAB-3(Rab3) with UNC-10(Rim) is required for docking to the plasma membrane to take place. AEX-3(RabGEF) activates both RAB-3(Rab3) and AEX-6(Rab27) although the role of AEX-6(Rab27) and its interacting proteins, CAB-1(NPCD1) and RBF-1 (Rabphilin), in control of neurotransmitter release is not clear. Studying docking in UNC-10(Rim) mutants using EM has been complicated by different results being obtained when using different methods of fixation (Gracheva *et al.*, 2008; Koushika *et al.*, 2001). This problem was not restricted to UNC-10(Rim) as the number of synaptic vesicles docked with the plasma membrane in UNC-13 (MUNC13) mutants as measured using EM varied under different fixation conditions and different definitions of docking (Gracheva *et al.*, 2006; Hammarlund *et al.*, 2007; Richmond *et al.*, 1999; Weimer *et al.*, 2006).

4. Endocytosis

Mutations in genes for synaptic vesicle endocytosis such as UNC-26(synaptoja-nin) (Harris *et al.*, 2000), EHS-1(Eps15), DYN-1(dynamin) (Salcini *et al.*, 2001), UNC-57(endophilin), UNC-11(AP180) (Nonet *et al.*, 1999), and UNC-41(stonin2) (Harada *et al.*, 1994; Jung *et al.*, 2007) cause decreased levels of ACh release. Mutations in many of these genes were isolated in the original screens for uncoordinated mutants and aldicarb resistant mutants (Brenner, 1974; Culotti *et al.*, 1981; Hosono *et al.*, 1987; Miller *et al.*, 1996; Nguyen *et al.*, 1995). Where tested these mutants were defective for the removal of synaptic vesicle membrane components such as SNB-1(synaptobrevin) from the plasma membrane demonstrating an endocytosis defect (Dittman and Kaplan, 2006). Mutations in the Ca^{2+} sensing protein SNT-1(synaptotagmin) also caused an endocytosis defect that resulted in a decrease in ACh release (Jorgensen et al., 1995; Nonet et al., 1993). This is different to mice where a knockout of synaptotagmin primarily caused a defect in exocytosis (Geppert et al., 1994). Recent experiments have also revealed that mutations in FAT-3 (Delta-6-desaturase) altered levels of polyunsaturated fatty acids leading to a mislocalization of UNC-26(synaptojanin) and thus a defect in endocytosis (Lesa et al., 2003; Marza et al., 2008; Watts et al., 2003).

5. Trafficking

UNC-104(KHC) is a kinesin heavy chain that forms part of a kinesin required to transport components of the synaptic vesicle cycle from the cell body to sites of neurotransmitter release, and mutations caused a strong reduction in ACh release (Bloom, 2001; Otsuka et al., 1991).

6. ACh-specific genes

So far all of the proteins in this section are thought to be involved in synaptic release of all neurotransmitters and are not specific to ACh release. There are a group of proteins that are specific for ACh release and these comprise some of the very first genes shown to be involved in C. elegans neurotransmission. CHA-1 (ChAT) mutants were the first mutants defective for ACh signaling to be identified (Rand and Russell, 1984). CHA-1(ChAT) encodes a choline acetyl-transferase required for the synthesis of ACh in the cytoplasm of cholinergic motorneurons (Alfonso et al., 1994b). ACh is then transported into synaptic vesicles via the UNC-17(vAChT) (Alfonso et al., 1993) vesicular ACh trans-porter. Interestingly, both the CHA-1(ChAT) and UNC-17(vAChT) genes are expressed from the same promoter ensuring that both are always present in cholinergic cells (Alfonso et al., 1994a). A mutation in a transmembrane domain of UNC-17(vAChT) was specifically suppressed by a mutation in the transmem-brane domain of SNB-1(synaptobrevin) (Sandoval et al., 2006). This evidence for an interaction between a neurotransmitter transporter and a SNARE protein is intriguing but the relevance of this interaction on ACh release is still unclear. The transport of ACh and other neurotransmitters into synaptic vesicles by their transporters requires a proton gradient in which the interior of the synaptic vesicle is acidified. Vacuolar H^+-ATPases are ATP-dependent proton pumps responsible for acidifying the synaptic vesicles. UNC-32(VoATPase) is expressed in motorneurons and encodes one component of the V-ATPase, although other homologous genes also exist in C. elegans (Pujol et al., 2001). Mutations in UNC-32(VoATPase) caused locomotion defects and an aldicarb resistance defect that suggested decreased ACh release (Brenner, 1974; Nguyen et al., 1995; Rand and Russell, 1985), although defects in V-ATPase are likely to affect release of many neurotransmitters, not just ACh. Once released from cholinergic motorneurons ACh must be rapidly degraded to prevent saturation of AChRs. Extracellular ACh is removed by AChEs, these are the enzymes inhibited by AChE inhibitors such as aldicarb. Analysis of acetylcholinesterase activity from crude extracts from 171 mutant strains identified the first acetyl-cholinesterase mutant, ACE-1 (Johnson et al., 1981). ACE-1 mutants had behaviors indistinguishable to that of wild type due to the presence of other acetylcholinesterases. An enhancer screen identified the ACE-2 mutation, which when combined with ACE-1 caused locomotion defects (Culotti et al., 1981). Analysis of ACE-1;ACE-2 double mutants revealed a third class of acetylcholinesterase (Kolson and Russell, 1985). Biochemical screening for defects in this third class of acetylcholinesterase yielded ACE-3 mutants (Johnson et al., 1988). ACE-1;ACE-3 and ACE-2;ACE-3 mutants had similar behaviors to wild-type animals. ACE-1;ACE-2;ACE-3 mutants were dead, sug-gesting that failure to remove extracellular ACh was lethal. Homology probing identified the ACE-1 gene (Arpagaus et al., 1994), which is expressed in the body-wall muscles. RT-PCR identified the ACE-2 and ACE-3 genes as well as a fourth acetylcholinesterase-like gene ACE-4 (Combes et al., 2000; Grauso et al., 1998). ACE-4 is transcribed in an intron with ACE-3 although it lacks critical residues required for acetylcholinesterase activity and is thought to be nonfunc-tional. AChEs convert ACh to acetic acid and choline. Choline can then be transported back into cells using the choline transporter CHO-1(CHT) (Okuda et al., 2000). CHO-1(CHT) null mutants were viable and caused a mild defect in ACh release under standard conditions (Matthies et al., 2006; Mullen et al., 2007). However, under conditions of strenuous muscle contraction the CHO-1(CHT) mutants did have strongly decreased muscle contractions compared to wild-type animals and this defect was reversed by addition of choline to the media. These results suggested that transport of choline is not essential for ACh synthesis as C. elegans appears to be able to synthesize choline de novo. The PMT-2(PMT) gene encodes phosphoethanolamine methyltrans-ferase, which is required for synthesis of phosphocholine (Palavalli et al., 2006). Null mutations in PMT-2(PMT) were larval lethal similar to CHA-1(ChAT) mutations. RNAi knockdown of PMT-2(PMT) caused a decrease in ACh release that was further decreased in the presence of the CHO-1(CHT) muta-tion, these effects of PMT-2(PMT) were relieved by addition of exogenous choline.

7. Essential genes as targets of regulatory genes

These results demonstrate that defects in synaptic vesicle cycling cause defects in signaling at the NMJ. As described earlier two genes essential for neurotransmitter release are targeted by the Gαq, Gα12, and Gαo pathways—UNC-13 and UNC-64(syntaxin) (Fig. 4.2). It is not clear whether other essential genes for neurotransmitter release are regulated by heterotrimeric G-protein signaling or by other signal transduction pathways. The targets of PKC-1(PKC) and KIN-1 (PKAc) are still unknown and perhaps they phosphorylate one or more of these proteins to control docking, exocytosis, or endocytosis of synaptic vesicles.

8. UNC-43(CAM kinase II)

Calmodulin protein kinase II (CAMKII) regulates synaptic strength in response to Ca^{2+} in mammalian neurons and is required for learning and memory in mice. The single C. elegans CAMKII UNC-43(CAMKII) is widely expressed in neurons and muscles (Reiner et al., 1999). UNC-43(CAMKII) gain-of-function mutants had severely reduced locomotion and this reduced locomotion was reversed by mutations in GOA-1(Gαo), DGK-1, EAT-16(GαqGAP), and GPB-2 (Reiner et al., 1999; Robatzek and Thomas, 2000). However, analysis of ACh signaling using electrophysiology of the body-wall muscles reported that UNC-43(CAMKII) gain-of-function mutants caused defects in ACh signaling that were not reversed by mutations in DGK-1 or EAT-16(GαqGAP). Instead, UNC-43(CAMKII) appeared to target the SLO-1 Ca^{2+} channel, probably by directly phosphorylating SLO-1. However, this report also failed to find any increase in ACh signaling by DGK-1 or EAT-16(GαqGAP) contrary to the experiments testing for sensitivity to aldicarb. It is possible that the ability of DGK-1 or EAT-16(GαqGAP) to suppress defects in UNC-43(CAMKII) gain-of-function mutants involves changes in GABA signaling onto the muscle, which was not tested.

9. Retrograde signaling from the body-wall muscles

Recent experiments have shown that the body-wall muscles signal to the motorneurons to alter levels of ACh release (Simon et al., 2008). Expression of miR-1, a muscle-specific microRNA, altered ACh signaling at the NMJ by decreasing the expression of some AChRs and the MEF-2 transcription factor in the muscle. Increases in MEF-2 expression in the muscles decreased ACh release onto the muscles. The molecular mechanism of this retrograde signal from the body-wall muscles to the motorneurons is unknown but it required RAB-3(Rab3), which regulates synaptic vesicle docking (see above). Loss of AEX-1, which has some limited homology to UNC-13, or loss of the prohormone convertase AEX-5 in muscles caused a defect in ACh release (Doi and Iwasaki, 2002). Defects in AEX-1 were suppressed by increasing EGL-30(Gαq) signaling, suggesting that AEX-1 and AEX-5 are required for retrograde signaling from muscles to motorneurons. Modulation of presynaptic activity by the postsynaptic cell is thought to be important to maintain the excitability of a synapse within a set range, this is often referred to as synaptic homeostasis (Macleod and Zinsmaier, 2006; O'Brien et al., 1998; Turrigiano et al., 1998). These results suggest that the cholinergic motorneurons are regulated both by neuromodulators and levels of activity of the postsynaptic cell, the body-wall muscle.

I. New screens—An embarrassment of riches

RNAi knockdown of genes is harder to achieve in the adult C. elegans nervous system than other cells, although in some mutant backgrounds a higher efficiency of gene knockdown in neurons is possible (Kennedy et al., 2004). These neuronal RNAi hypersensitive strains have been used to identify genes that either positively (Sieburth et al., 2005) or negatively (Vashlishan et al., 2008) regulate ACh signaling. These screens have identified at least 185 genes that modulate ACh signaling, less than half of which had been previously identified. New strategies have attempted to group mutants into phenotypic classes to determine the reasons why they are defective. One approach compares fluorescently labeled protein expression and localization to compare wild type with mutant animals. The labeled proteins used were ones known to be involved in different aspects of neuronal function such as exocytosis, endocytosis, and DCV release. (Ch'ng et al., 2008). If a mutation in a gene alters the localization of markers for endocytosis but not exocytosis then it is likely the gene being studied has a role in endocytosis. These techniques offer the hope of identifying yet more regulators of neuronal function and quickly identifying why the mutants are defective. Thus, the network of pathways that act to regulate ACh release are likely to grow and maintain the C. elegans NMJ as one of the most understood synapses in any animal. However, sorting these genes and identifying the pathways in which they operate will take many more years of work.

J. The C. elegans NMJ—A model for human synapses?

By EM C. elegans synapses appear very similar to vertebrate synapses (Fig. 4.1). However, vertebrate NMJs are typically thought to be "all or nothing" synapses where depolarization of the motorneuron reliably leads to release of enough ACh to always cause contraction of the muscle (Katz, 1969). In contrast, NMJs of invertebrates, such as C. elegans, appear to be graded synapses, where the level of depolarization can vary and levels of ACh release depends on how depolarized the motorneuron becomes and how many synaptic vesicles are docked to the membrane. Both depolarization and docking of synaptic vesicles can be modulated by the signaling pathways described here. Thus, the C. *elegans* NMJ displays some similarities to synapses in the mammalian nervous system, where depolarization of a presynaptic neuron does not always cause activation of the postsynaptic neuron. The signaling pathways that regulate C. *elegans* motorneurons are conserved in mammalian neurons: heterotrimeric G-protein signaling regulates neuronal function in mammalian neurons (Wettschureck and Offermanns, 2005) and RhoGTPases have been implicated in human mental retardation as well as learning and memory in rats (Diana *et al.*, 2007; Govek *et al.*, 2005). Both mammalian and C. *elegans* neurons use DAG activation of MUNC13 and PKC to regulate synaptic vesicle release (Brose and Rosenmund, 2002; Brose *et al.*, 1995; McMullan and Nurrish, 2007). Also, drugs used to treat human mental disorders have been shown to alter signaling at the C. *elegans* NMJ. For example, serotonin and drugs that act to modulate serotonin, for example, fluoxetine (Prozac), act to decrease DAG levels in cholinergic motorneurons (Nurrish *et al.*, 1999), although at high concentrations fluoxetine has effects on C. *elegans* that are independent of serotonin signaling (Choy and Thomas, 1999; Choy *et al.*, 2006; Ranganathan *et al.*, 2001). Another example is Valproate (VPA), which is used to treat epilepsy and bipolar disorder. In C. *elegans* VPA inhibited DAG-mediated ACh release (Tokuoka *et al.*, 2008). The effects of VPA appeared specific to DAG regulation of UNC-13 and not PKC-1 and the target of VPA appears to be on EGL-8(PLCβ) or one of its regulators. Heterotrimeric G-protein signaling pathways have also been implicated in the mechanism of action of volatile anesthetics as mutations in either the GOA-1(Gαo) or EGL-30(Gαq) pathway modulated changes in C. *elegans* response to human anesthetics such as halothane (Metz *et al.*, 2007; Morgan *et al.*, 2007; van Swinderen *et al.*, 1999, 2001).

Thus far all the components of the neuronal networks acting within the C. *elegans* motorneuron are conserved in humans (Fig. 4.2), and those orthologs that have been investigated in mammals have the same function in mammalian neurons as in C. *elegans* neurons (Brose *et al.*, 1995). It seems likely therefore that for future studies of intracellular networks in mammalian neurons the orthologs of the proteins that control the activity of the C. *elegans* NMJ would be a good place to start.

References

Afshar, K., Willard, F. S., Colombo, K., Johnston, C. A., McCudden, C. R., Siderovski, D. P., and Gonczy, P. (2004). RIC-8 is required for GPR-1/2-dependent Galpha function during asymmetric division of C. *elegans* embryos. Cell **119**(2), 219–230.

Alfonso, A., Grundahl, K., Duerr, J. S., Han, H. P., and Rand, J. B. (1993). The *Caenorhabditis elegans* unc-17 gene: A putative vesicular acetylcholine transporter. *Science* **261**(5121), 617–619.

Alfonso, A., Grundahl, K., McManus, J. R., Asbury, J. M., and Rand, J. B. (1994a). Alternative splicing leads to two cholinergic proteins in *Caenorhabditis elegans*. *J. Mol. Biol.* **241**(4), 627–630.

Alfonso, A., Grundahl, K., McManus, J. R., and Rand, J. B. (1994b). Cloning and characterization of the choline acetyltransferase structural gene (cha-1) from C. *elegans*. *J. Neurosci.* **14**(4), 2290–2300.

Ann, K., Kowalchyk, J. A., Loyet, K. M., and Martin, T. F. (1997). Novel Ca^{2+}-binding protein (CAPS) related to UNC-31 required for Ca^{2+}-activated exocytosis. *J. Biol. Chem.* **272**(32), 19637–19640.

Aravamudan, B., Fergestad, T., Davis, W. S., Rodesch, C. K., and Broadie, K. (1999). *Drosophila* UNC-13 is essential for synaptic transmission. *Nat. Neurosci.* **2**(11), 965–971.

Arpagaus, M., Fedon, Y., Cousin, X., Chatonnet, A., Berge, J. B., Fournier, D., and Toutant, J. P. (1994). cDNA sequence, gene structure, and *in vitro* expression of ace-1, the gene encoding acetylcholinesterase of class A in the nematode *Caenorhabditis elegans*. *J. Biol. Chem.* **269**(13), 9957–9965.

Augustin, I., Rosenmund, C., Sudhof, T. C., and Brose, N. (1999). Munc13-1 is essential for fusion competence of glutamatergic synaptic vesicles. *Nature* **400**(6743), 457–461.

Avery, L. (1993). The genetics of feeding in *Caenorhabditis elegans*. *Genetics* **133**(4), 897–917.

Avery, L., Bargmann, C. I., and Horvitz, H. R. (1993). The *Caenorhabditis elegans* unc-31 gene affects multiple nervous system-controlled functions. *Genetics* **134**(2), 455–464.

Bargmann, C. I. (1998). Neurobiology of the *Caenorhabditis elegans* genome. *Science* **282**, 2028–2033.

Bargmann, C. I. (2006). Chemosensation in C. *elegans*. *WormBook* 1–29.

Bargmann, C. I., and Kaplan, J. M. (1998). Signal transduction in the *Caenorhabditis elegans* nervous system. *Annu. Rev. Neurosci.* **21**, 279–308.

Barnes, T. M., and Hekimi, S. (1997). The *Caenorhabditis elegans* avermectin resistance and anesthetic response gene unc-9 encodes a member of a protein family implicated in electrical coupling of excitable cells. *J. Neurochem.* **69**(6), 2251–2260.

Bastiani, C., and Mendel, J. (2006). Heterotrimeric G proteins in C. *elegans*. *WormBook* 1–25.

Basu, J., Shen, N., Dulubova, I., Lu, J., Guan, R., Guryev, O., Grishin, N. V., Rosenmund, C., and Rizo, J. (2005). A minimal domain responsible for Munc13 activity. *Nat. Struct. Mol. Biol.* **12**(11), 1017–1018.

Basu, J., Betz, A., Brose, N., and Rosenmund, C. (2007). Munc13-1 C1 domain activation lowers the energy barrier for synaptic vesicle fusion. *J. Neurosci.* **27**(5), 1200–1210.

Berger, A. J., Hart, A. C., and Kaplan, J. M. (1998). G alphas-induced neurodegeneration in *Caenorhabditis elegans*. *J. Neurosci.* **18**(8), 2871–2880.

Betz, A., Okamoto, M., Benseler, F., and Brose, N. (1997). Direct interaction of the rat unc-13 homologue Munc13-1 with the N terminus of syntaxin. *J. Biol. Chem.* **272**(4), 2520–2526.

Betz, A., Thakur, P., Junge, H. J., Ashery, U., Rhee, J. S., Scheuss, V., Rosenmund, C., Rettig, J., and Brose, N. (2001). Functional interaction of the active zone proteins Munc13-1 and RIM1 in synaptic vesicle priming. *Neuron* **30**(1), 183–196.

Bloom, G. S. (2001). The UNC-104/KIF1 family of kinesins. *Curr. Opin. Cell Biol.* **13**(1), 36–40.

Brenner, S. (1974). The genetics of *Caenorhabditis elegans*. *Genetics* **77**(1), 71–94.

Brockie, P. J., and Maricq, A. V. (2006). Ionotropic glutamate receptors: Genetics, behavior and electrophysiology. *WormBook* 1–16.

Brose, N., and Rosenmund, C. (2002). Move over protein kinase C, you've got company: Alternative cellular effectors of diacylglycerol and phorbol esters. *J. Cell Sci.* **115**(Pt. 23), 4399–4411.

Brose, N., Hofmann, K., Hata, Y., and Sudhof, T. C. (1995). Mammalian homologues of *Caenorhabditis elegans* unc-13 gene define novel family of C2-domain proteins. *J. Biol. Chem.* **270**(42), 25273–25280.

Brown, L. A., Jones, A. K., Buckingham, S. D., Mee, C. J., and Sattelle, D. B. (2006). Contributions from *Caenorhabditis elegans* functional genetics to antiparasitic drug target identification and validation: Nicotinic acetylcholine receptors, a case study. *Int. J. Parasitol.* **36**(6), 617–624.

Brundage, L., Avery, L., Katz, A., Kim, U. J., Mendel, J. E., Sternberg, P. W., and Simon, M. I. (1996). Mutations in a C. *elegans* Gqalpha gene disrupt movement, egg laying, and viability. *Neuron* **16**(5), 999–1009.

Brunk, I., Blex, C., Speidel, D., Brose, N., and Ahnert-Hilger, G. (2009). Ca^{2+}-dependent activator proteins of secretion promote vesicular monoamine uptake. *J. Biol. Chem.* **284**, 1050–1056.

Carre-Pierrat, M., Baillie, D., Johnsen, R., Hyde, R., Hart, A., Granger, L., and Segalat, L. (2006). Characterization of the *Caenorhabditis elegans* G protein-coupled serotonin receptors. *Invert. Neurosci.* **6**(4), 189–205.

Charlie, N. K., Schade, M. A., Thomure, A. M., and Miller, K. G. (2006a). Presynaptic UNC-31 (CAPS) is required to activate the G alpha(s) pathway of the *Caenorhabditis elegans* synaptic signaling network. *Genetics* **172**(2), 943–961.

Charlie, N. K., Thomure, A. M., Schade, M. A., and Miller, K. G. (2006b). The Dunce cAMP phosphodiesterase PDE-4 negatively regulates G{alpha}s-dependent and G{alpha}s-independent cAMP pools in the C. *elegans* synaptic signaling network. *Genetics* **173**, 111–130.

Chase, D. L., Patikoglou, G. A., and Koelle, M. R. (2001). Two RGS proteins that inhibit Galpha(o) and Galpha(q) signaling in C. *elegans* neurons require a Gbeta(5)-like subunit for function. *Curr. Biol.* **11**(4), 222–231.

Chen, W., and Lim, L. (1994). The *Caenorhabditis elegans* small GTP-binding protein RhoA is enriched in the nerve ring and sensory neurons during larval development. *J. Biol. Chem.* **269**(51), 32394–32404.

Ch'ng, Q., Sieburth, D., and Kaplan, J. M. (2008). Profiling synaptic proteins identifies regulators of insulin secretion and lifespan. *PLoS Genet.* **4**(11), e1000283.

Choy, R. K., and Thomas, J. H. (1999). Fluoxetine-resistant mutants in C. *elegans* define a novel family of transmembrane proteins. *Mol. Cell* **4**(2), 143–152.

Choy, R. K., Kemner, J. M., and Thomas, J. H. (2006). Fluoxetine-resistance genes in *Caenorhabditis elegans* function in the intestine and may act in drug transport. *Genetics* **172**(2), 885–892.

Colquhoun, L., Holden-Dye, L., and Walker, R. J. (1991). The pharmacology of cholinoceptors on the somatic muscle cells of the parasitic nematode *Ascaris suum*. *J. Exp. Biol.* **158**, 509–530.

Combes, D., Fedon, Y., Grauso, M., Toutant, J. P., and Arpagaus, M. (2000). Four genes encode acetylcholinesterases in the nematodes *Caenorhabditis elegans* and *Caenorhabditis briggsae*. cDNA sequences, genomic structures, mutations and *in vivo* expression. *J. Mol. Biol.* **300**(4), 727–742.

Consortium, Ce S. (1998). Genome sequence of the nematode C. *elegans*: A platform for investigating biology. *Science* **282**, 2012–2018.

Coppola, T., Magnin-Luthi, S., Perret-Menoud, V., Gattesco, S., Schiavo, G., and Regazzi, R. (2001). Direct interaction of the Rab3 effector RIM with Ca^{2+} channels, SNAP-25, and synaptotagmin. *J. Biol. Chem.* **276**(35), 32756–32762.

Couwenbergs, C., Spilker, A. C., and Gotta, M. (2004). Control of embryonic spindle positioning and Galpha activity by C. *elegans* RIC-8. *Curr. Biol.* **14**(20), 1871–1876.

Culotti, J. G., Von Ehrenstein, G., Culotti, M. R., and Russell, R. L. (1981). A second class of acetylcholinesterase-deficient mutants of the nematode *Caenorhabditis elegans*. *Genetics* **97**(2), 281–305.

Dale, H. H. (1914). The action of certain esters and ethers of choline, and their relation to muscarine. *J. Pharmacol. Exp. Ther.* **6**, 147–190.

deBakker, C. D., Haney, L. B., Kinchen, J. M., Grimsley, C., Lu, M., Klingele, D., Hsu, P. K., Chou, B. K., Cheng, L. C., Blangy, A., Sondek, J., Hengartner, M. O., *et al.* (2004). Phagocytosis of apoptotic cells is regulated by a UNC-73/TRIO-MIG-2/rhog signaling module and armadillo repeats of CED-12/ELMO. *Curr. Biol.* **14**(24), 2208–2216.

Dernovici, S., Starc, T., Dent, J. A., and Ribeiro, P. (2007). The serotonin receptor SER-1 (5HT2ce) contributes to the regulation of locomotion in Caenorhabditis elegans. Dev. Neurobiol. 67(2), 189–204.

Diana, G., Valentini, G., Travaglione, S., Falzano, L., Pieri, M., Zona, C., Meschini, S., Fabbri, A., and Fiorentini, C. (2007). Enhancement of learning and memory after activation of cerebral Rho GTPases. Proc. Natl. Acad. Sci. USA 104(2), 636–641.

Dillon, C., and Goda, Y. (2005). The actin cytoskeleton: Integrating form and function at the synapse. Annu. Rev. Neurosci. 28, 25–55.

Dittman, J. S., and Kaplan, J. M. (2006). Factors regulating the abundance and localization of synaptobrevin in the plasma membrane. Proc. Natl. Acad. Sci. USA 103(30), 11399–11404.

Dittman, J. S., and Kaplan, J. M. (2008). Behavioral impact of neurotransmitter-activated G-protein-coupled receptors: Muscarinic and GABAB receptors regulate Caenorhabditis elegans locomotion. J. Neurosci. 28(28), 7104–7112.

Doi, M., and Iwasaki, K. (2002). Regulation of retrograde signaling at neuromuscular junctions by the novel C2 domain protein AEX-1. Neuron 33(2), 249–259.

Dolphin, A. C. (2006). A short history of voltage-gated calcium channels. Br. J. Pharmacol. 147 (Suppl. 1), S56–S62.

Dulubova, I., Sugita, S., Hill, S., Hosaka, M., Fernandez, I., Sudhof, T. C., and Rizo, J. (1999). A conformational switch in syntaxin during exocytosis: Role of munc18. EMBO J. 18(16), 4372–4382.

Dulubova, I., Lou, X., Lu, J., Huryeva, I., Alam, A., Schneggenburger, R., Sudhof, T. C., and Rizo, J. (2005). A Munc13/RIM/Rab3 tripartite complex: From priming to plasticity? EMBO J. 24(16), 2839–2850.

Dybbs, M., Ngai, J., and Kaplan, J. M. (2005). Using microarrays to facilitate positional cloning: Identification of tomosyn as an inhibitor of neurosecretion. PLoS Genet. 1(1), 6–16.

Elhamdani, A., Martin, T. F., Kowalchyk, J. A., and Artalejo, C. R. (1999). Ca(2+)-dependent activator protein for secretion is critical for the fusion of dense-core vesicles with the membrane in calf adrenal chromaffin cells. J. Neurosci. 19, 7375–7383.

Evans, G. J., and Morgan, A. (2003). Regulation of the exocytotic machinery by cAMP-dependent protein kinase: Implications for presynaptic plasticity. Biochem. Soc. Trans. 31(Pt. 4), 824–827.

Feng, Z., Li, W., Ward, A., Piggott, B. J., Larkspur, E. R., Sternberg, P. W., and Xu, X. Z. (2006). A C. elegans model of nicotine-dependent behavior: Regulation by trp-family channels. Cell 127(3), 621–633.

Fujita, Y., Shirataki, H., Sakisaka, T., Asakura, T., Ohya, T., Kotani, H., Yokoyama, S., Nishioka, H., Matsuura, Y., Mizoguchi, A., Scheller, R. H., and Takai, Y. (1998). Tomosyn: A syntaxin-1-binding protein that forms a novel complex in the neurotransmitter release process. Neuron 20(5), 905–915.

Fukuto, H. S., Ferkey, D. M., Apicella, A. J., Lans, H., Sharmeen, T., Chen, W., Lefkowitz, R. J., Jansen, G., Schafer, W. R., and Hart, A. C. (2004). G protein-coupled receptor kinase function is essential for chemosensation in C. elegans. Neuron 42(4), 581–593.

Gengyo-Ando, K., Kamiya, Y., Yamakawa, A., Kodaira, K., Nishiwaki, K., Miwa, J., Hori, I., and Hosono, R. (1993). The C. elegans unc-18 gene encodes a protein expressed in motor neurons. Neuron 11(4), 703–711.

Geppert, M., Goda, Y., Hammer, R. E., Li, C., Rosahl, T. W., Stevens, C. F., and Sudhof, T. C. (1994). Synaptotagmin I: A major Ca^{2+} sensor for transmitter release at a central synapse. Cell 79 (4), 717–727.

Giles, A. C., Rose, J. K., and Rankin, C. H. (2006). Investigations of learning and memory in Caenorhabditis elegans. Int. Rev. Neurobiol. 69, 37–71.

Goodman, M. B. (2006). Mechanosensation. WormBook 1–14.

Goto, K., Hozumi, Y., Nakano, T., Saino-Saito, S., and Martelli, A. M. (2008). Lipid messenger, diacylglycerol, and its regulator, diacylglycerol kinase, in cells, organs, and animals: History and perspective. *Tohoku J. Exp. Med.* **214**(3), 199–212.

Gotta, M., and Ahringer, J. (2001). Distinct roles for Galpha and Gbetagamma in regulating spindle position and orientation in *Caenorhabditis elegans* embryos. *Nat. Cell Biol.* **3**(3), 297–300.

Govek, E. E., Newey, S. E., and Van Aelst, L. (2005). The role of the Rho GTPases in neuronal development. *Genes Dev.* **19**(1), 1–49.

Gracheva, E. O., Burdina, A. O., Holgado, A. M., Berthelot-Grosjean, M., Ackley, B. D., Hadwiger, G., Nonet, M. L., Weimer, R. M., and Richmond, J. E. (2006). Tomosyn inhibits synaptic vesicle priming in *Caenorhabditis elegans*. *PLoS Biol.* **4**(8), e261.

Gracheva, E. O., Burdina, A. O., Touroutine, D., Berthelot-Grosjean, M., Parekh, H., and Richmond, J. E. (2007a). Tomosyn negatively regulates both synaptic transmitter and neuropeptide release at the C. *elegans* neuromuscular junction. *J. Physiol.* **585**, 705–709.

Gracheva, E. O., Burdina, A. O., Touroutine, D., Berthelot-Grosjean, M., Parekh, H., and Richmond, J. E. (2007b). Tomosyn negatively regulates CAPS-dependent peptide release at *Caenorhabditis elegans* synapses. *J. Neurosci.* **27**(38), 10176–10184.

Gracheva, E. O., Hadwiger, G., Nonet, M. L., and Richmond, J. E. (2008). Direct interactions between C. *elegans* RAB-3 and Rim provide a mechanism to target vesicles to the presynaptic density. *Neurosci. Lett.* **444**(2), 137–142.

Grauso, M., Culetto, E., Combes, D., Fedon, Y., Toutant, J. P., and Arpagaus, M. (1998). Existence of four acetylcholinesterase genes in the nematodes *Caenorhabditis elegans* and *Caenorhabditis briggsae*. *FEBS Lett.* **424**(3), 279–284.

Gross, R. E., Bagchi, S., Lu, X., and Rubin, C. S. (1990). Cloning, characterization, and expression of the gene for the catalytic subunit of camp-dependent protein kinase in *Caenorhabditis elegans*. Identification of highly conserved and unique isoforms generated by alternative splicing. *J. Biol. Chem.* **265**(12), 6896–6907.

Hajdu-Cronin, Y. M., Chen, W. J., Patikoglou, G., Koelle, M. R., and Sternberg, P. W. (1999). Antagonism between G(o)alpha and G(q)alpha in *Caenorhabditis elegans*: The RGS protein EAT-16 is necessary for G(o)alpha signaling and regulates G(q)alpha activity. *Genes Dev.* **13**(14), 1780–1793.

Hall, A. (1998). Rho GTPases and the actin cytoskeleton. *Science* **279**(5350), 509–514.

Hall, D. H. (2008). C. *elegans* Atlas. CSHL Press, New York.

Hammarlund, M., Palfreyman, M. T., Watanabe, S., Olsen, S., and Jorgensen, E. M. (2007). Open syntaxin docks synaptic vesicles. *PLoS Biol.* **5**(8), e198.

Hammarlund, M., Watanabe, S., Schuske, K., and Jorgensen, E. M. (2008). CAPS and syntaxin dock dense core vesicles to the plasma membrane in neurons. *J. Cell Biol.* **180**, 483–491.

Harada, S., Hori, I., Yamamoto, H., and Hosono, R. (1994). Mutations in the unc-41 gene cause elevation of acetylcholine levels. *J. Neurochem.* **63**(2), 439–446.

Hardie, R. C. (2003). Regulation of TRP channels via lipid second messengers. *Annu. Rev. Physiol.* **65**, 735–759.

Harris, T. W., Hartwieg, E., Horvitz, H. R., and Jorgensen, E. M. (2000). Mutations in synaptojanin disrupt synaptic vesicle recycling. *J. Cell Biol.* **150**(3), 589–600.

Hawasli, A. H., Saifee, O., Liu, C., Nonet, M. L., and Crowder, C. M. (2004). Resistance to volatile anesthetics by mutations enhancing excitatory neurotransmitter release in *Caenorhabditis elegans*. *Genetics* **168**(2), 831–843.

Hess, H. A., Roper, J. C., Grill, S. W., and Koelle, M. R. (2004). RGS-7 completes a receptor-independent heterotrimeric G protein cycle to asymmetrically regulate mitotic spindle positioning in C. *elegans*. *Cell* **119**(2), 209–218.

Hiley, E., McMullan, R., and Nurrish, S. J. (2006). The Galpha12-RGS RhoGEF–RhoA signalling pathway regulates neurotransmitter release in C. *elegans*. *EMBO J.* **25**, 5884–5895.

Hilfiker, S., Pieribone, A., Czernik, A. J., Kao, H. T., Augustine, G. J., and Greengard, P. (1999). Synapsins as regulators of neurotransmitter release. *Philos. Trans. R. Soc. Lond. B Biol. Sci.* **354** (1381), 269–279.

Hille, B. (2001). "Ion Channels of Excitable Membranes." Sinauer Associates, Sunderland, MA.

Hilliard, M. A., Apicella, A. J., Kerr, R., Suzuki, H., Bazzicalupo, P., and Schafer, W. R. (2005). *In vivo* imaging of C. *elegans* ASH neurons: Cellular response and adaptation to chemical repellents. *EMBO J.* **24**(1), 63–72.

Hobert, O. (2005). Specification of the nervous system. *WormBook* 1–19.

Hodgkin, A. L. (1964a). The conduction of the nervous impulse. Charles C. Thomas, Springfield, Ill., 108–120.

Hodgkin, A. L. (1964b). The ionic basis of nervous conduction. *Science* **145**, 1148–1154.

Hodgkin, A. L., and Huxley, A. F. (1952). A quantitative description of membrane current and its application to conduction and excitation in nerve. *J. Physiol.* **117**, 500–544.

Hosono, R., and Kamiya, Y. (1991). Additional genes which result in an elevation of acetylcholine levels by mutations in Caenorhabditis elegans. *Neurosci. Lett.* **128**(2), 243–244.

Hosono, R., Sassa, T., and Kuno, S. (1987). Mutations affecting acetylcholine levels in the nematode Caenorhabditis elegans. *J. Neurochem.* **49**(6), 1820–1823.

Hosono, R., Hekimi, S., Kamiya, Y., Sassa, T., Murakami, S., Nishiwaki, K., Miwa, J., Taketo, A., and Kodaira, K. I. (1992). The unc-18 gene encodes a novel protein affecting the kinetics of acetylcholine metabolism in the nematode Caenorhabditis elegans. *J. Neurochem.* **58**(4), 1517–1525.

Houssa, B., de Widt, J., Kranenburg, O., Moolenaar, W. H., and van Blitterswijk, W. J. (1999). Diacylglycerol kinase theta binds to and is negatively regulated by active RhoA. *J. Biol. Chem.* **274**(11), 6820–6822.

Husson, S. J., Janssen, T., Baggerman, G., Bogert, B., Kahn-Kirby, A. H., Ashrafi, K., and Schoofs, L. (2007). Impaired processing of FLP and NLP peptides in carboxypeptidase E (EGL-21)-deficient Caenorhabditis elegans as analyzed by mass spectrometry. *J. Neurochem.* **102**(1), 246–260.

Hwang, S. B., and Lee, J. (2003). Neuron cell type-specific SNAP-25 expression driven by multiple regulatory elements in the nematode Caenorhabditis elegans. *J. Mol. Biol.* **333**(2), 237–247.

Hwang, J. M., Chang, D. J., Kim, U. S., Lee, Y. S., Park, Y. S., Kaang, B. K., and Cho, N. J. (1999). Cloning and functional characterization of a Caenorhabditis elegans muscarinic acetylcholine receptor. *Recept. Channels* **6**(6), 415–424.

Iwasaki, K., and Toyonaga, R. (2000). The Rab3 GDP/GTP exchange factor homolog AEX-3 has a dual function in synaptic transmission. *EMBO J.* **19**(17), 4806–4816.

Iwasaki, K., Staunton, J., Saifee, O., Nonet, M., and Thomas, J. H. (1997). Aex-3 encodes a novel regulator of presynaptic activity in C. *elegans. Neuron* **18**(4), 613–622.

Jacob, T. C., and Kaplan, J. M. (2003). The EGL-21 carboxypeptidase E facilitates acetylcholine release at Caenorhabditis elegans neuromuscular junctions. *J. Neurosci.* **23**(6), 2122–2130.

Jansen, G., Thijssen, K. L., Werner, P., van der Horst, M., Hazendonk, E., and Plasterk, R. H. (1999). The complete family of genes encoding G proteins of Caenorhabditis elegans. *Nat. Genet.* **21**(4), 414–419.

Jansen, G., Weinkove, D., and Plasterk, R. H. (2002). The G-protein gamma subunit gpc-1 of the nematode C. *elegans* is involved in taste adaptation. *EMBO J.* **21**(5), 986–994.

Jin, Y. (2005). Synaptogenesis. *WormBook* 1–11.

Jockusch, W. J., Speidel, D., Sigler, A., Sorensen, J. B., Varoqueaux, F., Rhee, J. S., and Brose, N. (2007). CAPS-1 and CAPS-2 are essential synaptic vesicle priming proteins. *Cell* **131**, 796–808.

Johnson, C. D., Duckett, J. G., Culotti, J. G., Herman, R. K., Meneely, P. M., and Russell, R. L. (1981). An acetylcholinesterase-deficient mutant of the nematode Caenorhabditis elegans. *Genetics* **97**(2), 261–279.

Johnson, C. D., Rand, J. B., Herman, R. K., Stern, B. D., and Russell, R. L. (1988). The acetylcholinesterase genes of C. elegans: Identification of a third gene (ace-3) and mosaic mapping of a synthetic lethal phenotype. *Neuron* **1**(2), 165–173.

Johnson, J. R., Ferdek, P., Lian, L. Y., Barclay, J. W., Burgoyne, R. D., and Morgan, A. (2009). Binding of UNC-18 to the N-terminus of syntaxin is essential for neurotransmission in *Caenorhabditis elegans*. *Biochem. J.* **418**(1), 73–80.

Jones, A. K., and Sattelle, D. B. (2004). Functional genomics of the nicotinic acetylcholine receptor gene family of the nematode, *Caenorhabditis elegans*. *Bioessays* **26**(1), 39–49.

Jorgensen, E. M. (2005). Gaba. *WormBook* 1–13.

Jorgensen, E. M., Hartwieg, E., Schuske, K., Nonet, M. L., Jin, Y., and Horvitz, H. R. (1995). Defective recycling of synaptic vesicles in synaptotagmin mutants of *Caenorhabditis elegans*. *Nature* **378**(6553), 196–199.

Jung, N., Wienisch, M., Gu, M., Rand, J. B., Muller, S. L., Krause, G., Jorgensen, E. M., Klingauf, J., and Haucke, V. (2007). Molecular basis of synaptic vesicle cargo recognition by the endocytic sorting adaptor stonin 2. *J. Cell Biol.* **179**(7), 1497–1510.

Kaplan, J. M. (1996). Sensory signaling in *Caenorhabditis elegans*. *Curr. Opin. Neurobiol.* **6**, 494–499.

Karbowski, J., Schindelman, G., Cronin, C. J., Seah, A., and Sternberg, P. W. (2008). Systems level circuit model of C. *elegans* undulatory locomotion: Mathematical modeling and molecular genetics. *J. Comput. Neurosci.* **24**(3), 253–276.

Kass, J., Jacob, T. C., Kim, P., and Kaplan, J. M. (2001). The EGL-3 proprotein convertase regulates mechanosensory responses of *Caenorhabditis elegans*. *J. Neurosci.* **21**(23), 9265–9272.

Katz, B. (1966). "Nerve, Muscle, and Synapse" McGraw-Hill, New York.

Katz, B. (1969). "The Release of Neural Transmitter Substances." Liverpool University Press, Liverpool.

Keating, C. D., Kriek, N., Daniels, M., Ashcroft, N. R., Hopper, N. A., Siney, E. J., Holden-Dye, L., and Burke, J. F. (2003). Whole-genome analysis of 60 G protein-coupled receptors in *Caenorhabditis elegans* by gene knockout with RNAi. *Curr. Biol.* **13**(19), 1715–1720.

Kennedy, S., Wang, D., and Ruvkun, G. (2004). A conserved siRNA-degrading RNase negatively regulates RNA interference in C. *elegans*. *Nature* **427**(6975), 645–649.

Kimura, Y., Corcoran, E. E., Eto, K., Gengyo-Ando, K., Muramatsu, M. A., Kobayashi, R., Freedman, J. H., Mitani, S., Hagiwara, M., Means, A. R., and Tokumitsu, H. (2002). A CaMK cascade activates CRE-mediated transcription in neurons of *Caenorhabditis elegans*. *EMBO Rep.* **3**(10), 962–966.

Kiyonaka, S., Wakamori, M., Miki, T., Uriu, Y., Nonaka, M., Bito, H., Beedle, A. M., Mori, E., Hara, Y., De Waard, M., Kanagawa, M., Itakura, M., et al. (2007). RIM1 confers sustained activity and neurotransmitter vesicle anchoring to presynaptic Ca^{2+} channels. *Nat. Neurosci.* **10**(6), 691–701.

Klattenhoff, C., Montecino, M., Soto, X., Guzman, L., Romo, X., Garcia, M. A., Mellstrom, B., Naranjo, J. R., Hinrichs, M. V., and Olate, J. (2003). Human brain synembryn interacts with Gsalpha and Gqalpha and is translocated to the plasma membrane in response to isoproterenol and carbachol. *J. Cell Physiol.* **195**(2), 151–157.

Koelle, M. R., and Horvitz, H. R. (1996). EGL-10 regulates G protein signaling in the C. *elegans* nervous system and shares a conserved domain with many mammalian proteins. *Cell* **84**(1), 115–125.

Kohn, R. E., Duerr, J. S., McManus, J. R., Duke, A., Rakow, T. L., Maruyama, H., Moulder, G., Maruyama, N., Barstead, R. J., and Rand, J. B. (2000). Expression of multiple UNC-13 proteins in the *Caenorhabditis elegans* nervous system. *Mol. Biol. Cell* **11**(10), 3441–3452.

Kolson, D. L., and Russell, R. L. (1985). New acetylcholinesterase-deficient mutants of the nematode *Caenorhabditis elegans*. *J. Neurogenet.* **2**(2), 69–91.

Korswagen, H. C., Park, J. H., Ohshima, Y., and Plasterk, R. H. (1997). An activating mutation in a *Caenorhabditis elegans* Gs protein induces neural degeneration. *Genes Dev.* **11**(12), 1493–1503.

Koushika, S. P., Richmond, J. E., Hadwiger, G., Weimer, R. M., Jorgensen, E. M., and Nonet, M. L. (2001). A post-docking role for active zone protein Rim. *Nat. Neurosci.* **4**(10), 997–1005.

Kubiak, T. M., Larsen, M. J., Zantello, M. R., Bowman, J. W., Nulf, S. C., and Lowery, D. E. (2003). Functional annotation of the putative orphan *Caenorhabditis elegans* G-protein-coupled receptor C10C6.2 as a FLP15 peptide receptor. *J. Biol. Chem.* **278**(43), 42115–42120.

Lackner, M. R., Nurrish, S. J., and Kaplan, J. M. (1999). Facilitation of synaptic transmission by EGL-30 Gqalpha and EGL-8 plcbeta: DAG binding to UNC-13 is required to stimulate acetylcholine release. *Neuron* **24**(2), 335–346.

Lee, C. H., Park, D., Wu, D., Rhee, S. G., and Simon, M. I. (1992). Members of the Gq alpha subunit gene family activate phospholipase C beta isozymes. *J. Biol. Chem.* **267**(23), 16044–16047.

Lee, R. Y., Sawin, E. R., Chalfie, M., Horvitz, H. R., and Avery, L. (1999a). EAT-4, a homolog of a mammalian sodium-dependent inorganic phosphate cotransporter, is necessary for glutamatergic neurotransmission in *Caenorhabditis elegans*. *J. Neurosci.* **19**(1), 159–167.

Lee, Y. S., Park, Y. S., Chang, D. J., Hwang, J. M., Min, C. K., Kaang, B. K., and Cho, N. J. (1999b). Cloning and expression of a G protein-linked acetylcholine receptor from *Caenorhabditis elegans*. *J. Neurochem.* **72**(1), 58–65.

Lee, Y. S., Park, Y. S., Nam, S., Suh, S. J., Lee, J., Kaang, B. K., and Cho, N. J. (2000). Characterization of GAR-2, a novel G protein-linked acetylcholine receptor from *Caenorhabditis elegans*. *J. Neurochem.* **75**(5), 1800–1809.

Lesa, G. M., Palfreyman, M., Hall, D. H., Clandinin, M. T., Rudolph, C., Jorgensen, E. M., and Schiavo, G. (2003). Long chain polyunsaturated fatty acids are required for efficient neurotransmission in *C. elegans*. *J. Cell Sci.* **116**(Pt. 24), 4965–4975.

Lewis, J. A., Wu, C. H., Berg, H., and Levine, J. H. (1980a). The genetics of levamisole resistance in the nematode *Caenorhabditis elegans*. *Genetics* **95**(4), 905–928.

Lewis, J. A., Wu, C. H., Levine, J. H., and Berg, H. (1980b). Levamisole-resistant mutants of the nematode *Caenorhabditis elegans* appear to lack pharmacological acetylcholine receptors. *Neuroscience* **5**(6), 967–989.

Li, C., and Kim, K. (2008). Neuropeptides. *WormBook* 1–36.

Liewald, J. F., Brauner, M., Stephens, G. J., Bouhours, M., Schultheis, C., Zhen, M., and Gottschalk, A. (2008). Optogenetic analysis of synaptic function. *Nat. Methods* **5**, 895–902.

Liu, Q., Chen, B., Ge, Q., and Wang, Z. W. (2007). Presynaptic Ca^{2+}/calmodulin-dependent protein kinase II modulates neurotransmitter release by activating BK channels at *Caenorhabditis elegans* neuromuscular junction. *J. Neurosci.* **27**(39), 10404–10413.

Liu, Y., Schirra, C., Stevens, D. R., Matti, U., Speidel, D., Hof, D., Bruns, D., Brose, N., and Rettig, J. (2008). CAPS facilitates filling of the rapidly releasable pool of large dense-core vesicles. *J. Neurosci.* **28**, 5594–5601.

Loewi, O. (1921). Uber hormonale iibertragbarkeit der Herznervenwirkung. *Pflügers Archiv* **189**, 239.

Lou, X., Korogod, N., Brose, N., and Schneggenburger, R. (2008). Phorbol esters modulate spontaneous and Ca^{2+}-evoked transmitter release via acting on both Munc13 and protein kinase C. *J. Neurosci.* **28**(33), 8257–8267.

Macleod, G. T., and Zinsmaier, K. E. (2006). Synaptic homeostasis on the fast track. *Neuron* **52**(4), 569–571.

Madison, J. M., Nurrish, S., and Kaplan, J. M. (2005). UNC-13 interaction with syntaxin is required for synaptic transmission. *Curr. Biol.* **15**(24), 2236–2242.

Mahoney, T. R., Liu, Q., Itoh, T., Luo, S., Hadwiger, G., Vincent, R., Wang, Z. W., Fukuda, M., and Nonet, M. L. (2006). Regulation of synaptic transmission by RAB-3 and RAB-27 in *Caenorhabditis elegans*. *Mol. Biol. Cell* **17**, 2617–2625.

Majewski, H., and Iannazzo, L. (1998). Protein kinase C: A physiological mediator of enhanced transmitter output. *Prog. Neurobiol.* **55**(5), 463–475.

Malenka, R. C., Madison, D. V., and Nicoll, R. A. (1986). Potentiation of synaptic transmission in the hippocampus by phorbol esters. *Nature* **321**(6066), 175–177.

Maruyama, N., and Brenner, S. (1991). A phorbol ester/diacylglycerol-binding protein encoded by the unc-13 gene of *Caenorhabditis elegans*. *Proc. Natl. Acad. Sci. USA* **88**(13), 5729–5733.

Marza, E., Long, T., Saiardi, A., Sumakovic, M., Eimer, S., Hall, D. H., and Lesa, G. M. (2008). Polyunsaturated fatty acids influence synaptojanin localization to regulate synaptic vesicle recycling. *Mol. Biol. Cell* **19**(3), 833–842.

Matthies, D. S., Fleming, P. A., Wilkes, D. M., and Blakely, R. D. (2006). The *Caenorhabditis elegans* choline transporter CHO-1 sustains acetylcholine synthesis and motor function in an activity-dependent manner. *J. Neurosci.* **26**(23), 6200–6212.

McEwen, J. M., and Kaplan, J. M. (2008). UNC-18 promotes both the anterograde trafficking and synaptic function of syntaxin. *Mol. Biol. Cell* **19**(9), 3836–3846.

McEwen, J. M., Madison, J. M., Dybbs, M., and Kaplan, J. M. (2006). Antagonistic regulation of synaptic vesicle priming by Tomosyn and UNC-13. *Neuron* **51**(3), 303–315.

McMullan, R., and Nurrish, S. J. (2007). Rho deep in thought. *Genes Dev.* **21**(21), 2677–2682.

McMullan, R., Hiley, E., Morrison, P., and Nurrish, S. J. (2006). Rho is a presynaptic activator of neurotransmitter release at pre-existing synapses in C. *elegans*. *Genes Dev.* **20**(1), 65–76.

Mendel, J. E., Korswagen, H. C., Liu, K. S., Hajdu-Cronin, Y. M., Simon, M. I., Plasterk, R. H., and Sternberg, P. W. (1995). Participation of the protein Go in multiple aspects of behavior in C. *elegans*. *Science* **267**(5204), 1652–1655.

Merida, I., Avila-Flores, A., and Merino, E. (2008). Diacylglycerol kinases: At the hub of cell signalling. *Biochem. J.* **409**(1), 1–18.

Metz, L. B., Dasgupta, N., Liu, C., Hunt, S. J., and Crowder, C. M. (2007). An evolutionarily conserved presynaptic protein is required for isoflurane sensitivity in *Caenorhabditis elegans*. *Anesthesiology* **107**(6), 971–982.

Miller, K. G., and Rand, J. B. (2000). A role for RIC-8 (Synembryn) and GOA-1 (G(o)alpha) in regulating a subset of centrosome movements during early embryogenesis in *Caenorhabditis elegans*. *Genetics* **156**(4), 1649–1660.

Miller, K. G., Alfonso, A., Nguyen, M., Crowell, J. A., Johnson, C. D., and Rand, J. B. (1996). A genetic selection for *Caenorhabditis elegans* synaptic transmission mutants. *Proc. Natl. Acad. Sci. USA* **93**(22), 12593–12598.

Miller, K. G., Emerson, M. D., and Rand, J. B. (1999). Goalpha and diacylglycerol kinase negatively regulate the Gqalpha pathway in C. *elegans*. *Neuron* **24**(2), 323–333.

Miller, K. G., Emerson, M. D., McManus, J. R., and Rand, J. B. (2000). RIC-8 (Synembryn): A novel conserved protein that is required for G(q)alpha signaling in the C. *elegans* nervous system. *Neuron* **27**(2), 289–299.

Morgan, P. G., Kayser, E. B., and Sedensky, M. M. (2007). C. *elegans* and volatile anesthetics. *WormBook* 1–11.

Mori, I., Sasakura, H., and Kuhara, A. (2007). Worm thermotaxis: A model system for analyzing thermosensation and neural plasticity. *Curr. Opin. Neurobiol.* **17**, 712–719.

Mullen, G. P., Mathews, E. A., Vu, M. H., Hunter, J. W., Frisby, D. L., Duke, A., Grundahl, K., Osborne, J. D., Crowell, J. A., and Rand, J. B. (2007). Choline transport and de novo choline synthesis support acetylcholine biosynthesis in *Caenorhabditis elegans* cholinergic neurons. *Genetics* **177**(1), 195–204.

Murray, P., Clegg, R. A., Rees, H. H., and Fisher, M. J. (2008). siRNA-mediated knockdown of a splice variant of the PK-A catalytic subunit gene causes adult-onset paralysis in *C. elegans*. *Gene* **408**(1–2), 157–163.

Nagel, G., Brauner, M., Liewald, J. F., Adeishvili, N., Bamberg, E., and Gottschalk, A. (2005). Light activation of channelrhodopsin-2 in excitable cells of *Caenorhabditis elegans* triggers rapid behavioral responses. *Curr. Biol.* **15**, 2279–2284.

Nelson, L. S., Rosoff, M. L., and Li, C. (1998). Disruption of a neuropeptide gene, flp-1, causes multiple behavioral defects in *Caenorhabditis elegans*. *Science* **281**(5383), 1686–1690.

Nguyen, P. V., and Woo, N. H. (2003). Regulation of hippocampal synaptic plasticity by cyclic AMP-dependent protein kinases. *Prog. Neurobiol.* **71**(6), 401–437.

Nguyen, M., Alfonso, A., Johnson, C. D., and Rand, J. B. (1995). *Caenorhabditis elegans* mutants resistant to inhibitors of acetylcholinesterase. *Genetics* **140**(2), 527–535.

Nonet, M. L., Grundahl, K., Meyer, B. J., and Rand, J. B. (1993). Synaptic function is impaired but not eliminated in *C. elegans* mutants lacking synaptotagmin. *Cell* **73**(7), 1291–1305.

Nonet, M. L., Staunton, J. E., Kilgard, M. P., Fergestad, T., Hartwieg, E., Horvitz, H. R., Jorgensen, E. M., and Meyer, B. J. (1997). *Caenorhabditis elegans* rab-3 mutant synapses exhibit impaired function and are partially depleted of vesicles. *J. Neurosci.* **17**(21), 8061–8073.

Nonet, M. L., Saifee, O., Zhao, H., Rand, J. B., and Wei, L. (1998). Synaptic transmission deficits in *Caenorhabditis elegans* synaptobrevin mutants. *J. Neurosci.* **18**(1), 70–80.

Nonet, M. L., Holgado, A. M., Brewer, F., Serpe, C. J., Norbeck, B. A., Holleran, J., Wei, L., Hartwieg, E., Jorgensen, E. M., and Alfonso, A. (1999). UNC-11, a *Caenorhabditis elegans* AP180 homologue, regulates the size and protein composition of synaptic vesicles. *Mol. Biol. Cell* **10**(7), 2343–2360.

Nurrish, S. J. (2002). An overview of *C. elegans* trafficking mutants. *Traffic* **3**(1), 2–10.

Nurrish, S., Segalat, L., and Kaplan, J. M. (1999). Serotonin inhibition of synaptic transmission: Galpha(0) decreases the abundance of UNC-13 at release sites. *Neuron* **24**(1), 231–242.

O'Brien, R. J., Kamboj, S., Ehlers, M. D., Rosen, K. R., Fischbach, G. D., and Huganir, R. L. (1998). Activity-dependent modulation of synaptic AMPA receptor accumulation. *Neuron* **21**(5), 1067–1078.

Okuda, T., Haga, T., Kanai, Y., Endou, H., Ishihara, T., and Katsura, I. (2000). Identification and characterization of the high-affinity choline transporter. *Nat. Neurosci.* **3**(2), 120–125.

Otsuka, A. J., Jeyaprakash, A., Garcia-Anoveros, J., Tang, L. Z., Fisk, G., Hartshorne, T., Franco, R., and Born, T. (1991). The *C. elegans* unc-104 gene encodes a putative kinesin heavy chain-like protein. *Neuron* **6**(1), 113–122.

Palavalli, L. H., Brendza, K. M., Haakenson, W., Cahoon, R. E., McLaird, M., Hicks, L. M., McCarter, J. P., Williams, D. J., Hresko, M. C., and Jez, J. M. (2006). Defining the role of phosphomethylethanolamine N-methyltransferase from *Caenorhabditis elegans* in phosphocholine biosynthesis by biochemical and kinetic analysis. *Biochemistry* **45**(19), 6056–6065.

Palmitessa, A., Hess, H. A., Bany, A., Kim, Y. M., Koelle, M. R., and Benovic, J. L. (2005). *Caenorhabditus elegans* arrestin regulates neural G protein signaling and olfactory adaptation and recovery. *J. Biol. Chem.* **280**(26), 24649–24662.

Papaioannou, S., Marsden, D., Franks, C. J., Walker, R. J., and Holden-Dye, L. (2005). Role of a FMRFamide-like family of neuropeptides in the pharyngeal nervous system of *Caenorhabditis elegans*. *J. Neurobiol.* **65**, 304–319.

Papaioannou, S., Holden-Dye, L., and Walker, R. J. (2008). The actions of *Caenorhabditis elegans* neuropeptide-like peptides (NLPs) on body wall muscle of *Ascaris suum* and pharyngeal muscle of *C. elegans*. *Acta Biol. Hung.* **59**(Suppl.), 189–197.

Park, Y. S., Lee, Y. S., Cho, N. J., and Kaang, B. K. (2000). Alternative splicing of gar-1, a *Caenorhabditis elegans* G-protein-linked acetylcholine receptor gene. *Biochem. Biophys. Res. Commun.* **268**(2), 354–358.

Phelan, P. (2005). Innexins: Members of an evolutionarily conserved family of gap-junction proteins. *Biochim. Biophys. Acta* **1711**(2), 225–245.

Pujol, N., Bonnerot, C., Ewbank, J. J., Kohara, Y., and Thierry-Mieg, D. (2001). The *Caenorhabditis elegans* unc-32 gene encodes alternative forms of a vacuolar ATPase a subunit. *J. Biol. Chem.* **276** (15), 11913–11921.

Rand, J. B. (2007). Acetylcholine. *WormBook* 1–21.

Rand, J. B., and Johnson, C. D. (1995). Genetic pharmacology: Interactions between drugs and gene products in *Caenorhabditis elegans*. *Methods Cell Biol.* **48**, 187–204.

Rand, J. B., and Russell, R. L. (1984). Choline acetyltransferase-deficient mutants of the nematode *Caenorhabditis elegans*. *Genetics* **106**(2), 227–248.

Rand, J. B., and Russell, R. L. (1985). Molecular basis of drug-resistance mutations in *C. elegans*. *Psychopharmacol. Bull.* **21**(3), 623–630.

Rand, J. B., Duerr, J. S., and Frisby, D. L. (1998). Neurogenetics of synaptic transmission in *Caenorhabditis elegans*. *Adv. Pharmacol.* **42**, 940–944.

Ranganathan, R., Cannon, S. C., and Horvitz, H. R. (2000). MOD-1 is a serotonin-gated chloride channel that modulates locomotory behaviour in *C. elegans*. *Nature* **408**(6811), 470–475.

Ranganathan, R., Sawin, E. R., Trent, C., and Horvitz, H. R. (2001). Mutations in the *Caenorhabditis elegans* serotonin reuptake transporter MOD-5 reveal serotonin-dependent and -independent activities of fluoxetine. *J. Neurosci.* **21**(16), 5871–5884.

Reiner, D. J., Newton, E. M., Tian, H., and Thomas, J. H. (1999). Diverse behavioural defects caused by mutations in *Caenorhabditis elegans* unc-43 cam kinase II. *Nature* **402**(6758), 199–203.

Reynolds, N. K., Schade, M. A., and Miller, K. G. (2005). Convergent, RIC-8-dependent Galpha signaling pathways in the *Caenorhabditis elegans* synaptic signaling network. *Genetics* **169**(2), 651–670.

Rhee, J. S., Betz, A., Pyott, S., Reim, K., Varoqueaux, F., Augustin, I., Hesse, D., Sudhof, T. C., Takahashi, M., Rosenmund, C., and Brose, N. (2002). Beta phorbol ester- and diacylglycerol-induced augmentation of transmitter release is mediated by Munc13s and not by PKCs. *Cell* **108** (1), 121–133.

Richmond, J. (2005). Synaptic function. *WormBook* 1–14.

Richmond, J. E. (2006). Electrophysiological recordings from the neuromuscular junction of *C. elegans*. *WormBook* 1–8.

Richmond, J. E., and Broadie, K. S. (2002). The synaptic vesicle cycle: Exocytosis and endocytosis in *Drosophila* and *C. elegans*. *Curr. Opin. Neurobiol.* **12**(5), 499–507.

Richmond, J. E., and Jorgensen, E. M. (1999). One GABA and two acetylcholine receptors function at the *C. elegans* neuromuscular junction. *Nat. Neurosci.* **2**(9), 791–797.

Richmond, J. E., Davis, W. S., and Jorgensen, E. M. (1999). UNC-13 is required for synaptic vesicle fusion in *C. elegans*. *Nat. Neurosci.* **2**(11), 959–964.

Richmond, J. E., Weimer, R. M., and Jorgensen, E. M. (2001). An open form of syntaxin bypasses the requirement for UNC-13 in vesicle priming. *Nature* **412**(6844), 338–341.

Robatzek, M., and Thomas, J. H. (2000). Calcium/calmodulin-dependent protein kinase II regulates *Caenorhabditis elegans* locomotion in concert with a G(o)/G(q) signaling network. *Genetics* **156**(3), 1069–1082.

Robatzek, M., Niacaris, T., Steger, K., Avery, L., and Thomas, J. H. (2001). Eat-11 encodes GPB-2, a Gbeta(5) ortholog that interacts with G(o)alpha and G(q)alpha to regulate *C. elegans* behavior. *Curr. Biol.* **11**(4), 288–293.

Rogers, C. M., Franks, C. J., Walker, R. J., Burke, J. F., and Holden-Dye, L. (2001). Regulation of the pharynx of *Caenorhabditis elegans* by 5-HT, octopamine, and FMRFamide-like neuropeptides. *J. Neurobiol.* **49**, 235–244.

Romo, X., Pasten, P., Martinez, S., Soto, X., Lara, P., de Arellano, A. R., Torrejon, M., Montecino, M., Hinrichs, M. V., and Olate, J. (2008). Xric-8 is a GEF for Gsalpha and partici-pates in maintaining meiotic arrest in *Xenopus laevis* oocytes. *J. Cell Physiol.* 214(3), 673–680.

Rosoff, M. L., Doble, K. E., Price, D. A., and Li, C. (1993). The flp-1 propeptide is processed into multiple, highly similar FMRFamide-like peptides in *Caenorhabditis elegans*. *Peptides* 14(2), 331–338.

Ross, E. M., and Wilkie, T. M. (2000). GTPase-activating proteins for heterotrimeric G proteins: Regulators of G protein signaling (RGS) and RGS-like proteins. *Annu. Rev. Biochem.* 69, 795–827.

Roth, M. G. (2008). Molecular mechanisms of PLD function in membrane traffic. *Traffic* 9(8), 1233–1239.

Saifee, O., Wei, L., and Nonet, M. L. (1998). The *Caenorhabditis elegans* unc-64 locus encodes a syntaxin that interacts genetically with synaptobrevin. *Mol. Biol. Cell* 9(6), 1235–1252.

Sakisaka, T., Baba, T., Tanaka, S., Izumi, G., Yasumi, M., and Takai, Y. (2004). Regulation of SNAREs by tomosyn and ROCK: Implication in extension and retraction of neurites. *J. Cell Biol.* 166(1), 17–25.

Salcini, A. E., Hilliard, M. A., Croce, A., Arbucci, S., Luzzi, P., Tacchetti, C., Daniell, L., De Camilli, P., Pelicci, P. G., Di Fiore, P. P., and Bazzicalupo, P. (2001). The Eps15 C. *elegans* homologue EHS-1 is implicated in synaptic vesicle recycling. *Nat. Cell Biol.* 3(8), 755–760.

Sandoval, G. M., Duerr, J. S., Hodgkin, J., Rand, J. B., and Ruvkun, G. (2006). A genetic interaction between the vesicular acetylcholine transporter VAChT/UNC-17 and synaptobrevin/SNB-1 in C. *elegans*. *Nat. Neurosci.* 9(5), 599–601.

Sassa, T., Ogawa, H., Kimoto, M., and Hosono, R. (1996). The synaptic protein UNC-18 is phosphorylated by protein kinase C. *Neurochem. Int.* 29(5), 543–552.

Sassa, T., Harada, S., Ogawa, H., Rand, J. B., Maruyama, N., and Hosono, R. (1999). Regulation of the UNC-18–*Caenorhabditis elegans* syntaxin complex by UNC-13. *J. Neurosci.* 19(12), 4772–4777.

Sattelle, D. B., and Buckingham, S. D. (2006). Invertebrate studies and their ongoing contributions to neuroscience. *Invert. Neurosci.* 6, 1–3.

Sawin, E. R., Ranganathan, R., and Horvitz, H. R. (2000). C. *elegans* locomotory rate is modulated by the environment through a dopaminergic pathway and by experience through a serotonergic pathway. *Neuron* 26(3), 619–631.

Schade, M. A., Reynolds, N. K., Dollins, C. M., and Miller, K. G. (2005). Mutations that rescue the paralysis of *Caenorhabditis elegans* ric-8 (synembryn) mutants activate the G alpha(s) pathway and define a third major branch of the synaptic signaling network. *Genetics* 169(2), 631–649.

Schafer, W. R., and Kenyon, C. J. (1995). A calcium-channel homologue required for adaptation to dopamine and serotonin in *Caenorhabditis elegans*. *Nature* 375(6526), 73–78.

Schafer, W. R., Sanchez, B. M., and Kenyon, C. J. (1996). Genes affecting sensitivity to serotonin in *Caenorhabditis elegans*. *Genetics* 143(3), 1219–1230.

Schoch, S., Castillo, P. E., Jo, T., Mukherjee, K., Geppert, M., Wang, Y., Schmitz, F., Malenka, R. C., and Sudhof, T. C. (2002). RIM1alpha forms a protein scaffold for regulating neurotransmitter release at the active zone. *Nature* 415(6869), 321–326.

Segalat, L., Elkes, D. A., and Kaplan, J. M. (1995). Modulation of serotonin-controlled behaviors by Go in *Caenorhabditis elegans*. *Science* 267(5204), 1648–1651.

Siddiqui, S. S., and Culotti, J. G. (1991). Examination of neurons in wild type and mutants of *Caenorhabditis elegans* using antibodies to horseradish peroxidase. *J. Neurogenet.* 7(4), 193–211.

Siderovski, D. P., and Willard, F. S. (2005). The GAPs, GEFs, and GDIs of heterotrimeric G-protein alpha subunits. *Int. J. Biol. Sci.* 1(2), 51–66.

Sieburth, D., Ch'ng, Q., Dybbs, M., Tavazoie, M., Kennedy, S., Wang, D., Dupuy, D., Rual, J. F., Hill, D. E., Vidal, M., Ruvkun, G., and Kaplan, J. M. (2005). Systematic analysis of genes required for synapse structure and function. *Nature* **436**(7050), 510–517.

Sieburth, D., Madison, J. M., and Kaplan, J. M. (2006). PKC-1 regulates secretion of neuropeptides. *Nat. Neurosci.* **10**, 49–57.

Sieburth, D., Madison, J. M., and Kaplan, J. M. (2007). PKC-1 regulates secretion of neuropeptides. *Nat. Neurosci.* **10**(1), 49–57.

Simon, D. J., Madison, J. M., Conery, A. L., Thompson-Peer, K. L., Soskis, M., Ruvkun, G. B., Kaplan, J. M., and Kim, J. K. (2008). The microRNA miR-1 regulates a MEF-2-dependent retrograde signal at neuromuscular junctions. *Cell* **133**(5), 903–915.

Speese, S., Petrie, M., Schuske, K., Ailion, M., Ann, K., Iwasaki, K., Jorgensen, E. M., and Martin, T. F. (2007). UNC-31 (CAPS) is required for dense-core vesicle but not synaptic vesicle exocytosis in *Caenorhabditis elegans*. *J. Neurosci.* **27**(23), 6150–6162.

Speidel, D., Bruederle, C. E., Enk, C., Voets, T., Varoqueaux, F., Reim, K., Becherer, U., Fornai, F., Ruggieri, S., Holighaus, Y., Weihe, E., Bruns, D., *et al.* (2005). CAPS1 regulates catecholamine loading of large densecore vesicles. *Neuron* **46**, 75–88.

Speidel, D., Salehi, A., Obermueller, S., Lundquist, I., Brose, N., Renstrom, E., and Rorsman, P. (2008). CAPS1 and CAPS2 regulate stability and recruitment of insulin granules in mouse pancreatic beta cells. *Cell Metab.* **7**, 57–67.

Steger, K. A., and Avery, L. (2004). The GAR-3 muscarinic receptor cooperates with calcium signals to regulate muscle contraction in the *Caenorhabditis elegans* pharynx. *Genetics* **167**(2), 633–643.

Steven, R., Kubiseski, T. J., Zheng, H., Kulkarni, S., Mancillas, J., Ruiz Morales, A., Hogue, C. W., Pawson, T., and Culotti, J. (1998). UNC-73 activates the Rac GTPase and is required for cell and growth cone migrations in *C. elegans*. *Cell* **92**(6), 785–795.

Steven, R., Zhang, L., Culotti, J., and Pawson, T. (2005). The UNC-73/Trio RhoGEF-2 domain is required in separate isoforms for the regulation of pharynx pumping and normal neurotransmission in *C. elegans*. *Genes Dev.* **19**(17), 2016–2029.

Stevens, D. R., Wu, Z. X., Matti, U., Junge, H. J., Schirra, C., Becherer, U., Wojcik, S. M., Brose, N., and Rettig, J. (2005). Identification of the minimal protein domain required for priming activity of Munc13-1. *Curr. Biol.* **15**(24), 2243–2248.

Stretton, A., Donmoyer, J., Davis, R., Meade, J., Cowden, C., and Sithigorngul, P. (1992). Motor behavior and motor nervous system function in the nematode *Ascaris suum*. *J. Parasitol.* **78**, 206–214.

Sudhof, T. C., and Rothman, J. E. (2009). Membrane fusion: Grappling with SNARE and SM proteins. *Science* **323**(5913), 474–477.

Suh, S. J., Park, Y. S., Lee, Y. S., Cho, T. J., Kaang, B. K., and Cho, N. J. (2001). Three functional isoforms of GAR-2, a *Caenorhabditis elegans* G-protein-linked acetylcholine receptor, are produced by alternative splicing. *Biochem. Biophys. Res. Commun.* **288**(5), 1238–1243.

Tabish, M., Clegg, R. A., Rees, H. H., and Fisher, M. J. (1999). Organization and alternative splicing of the *Caenorhabditis elegans* cAMP-dependent protein kinase catalytic-subunit gene (kin-1). *Biochem. J.* **339**(Pt. 1), 209–216.

Tabuse, Y., Nishiwaki, K., and Miwa, J. (1989). Mutations in a protein kinase C homolog confer phorbol ester resistance on *Caenorhabditis elegans*. *Science* **243**(4899), 1713–1716.

Tall, G. G., Krumins, A. M., and Gilman, A. G. (2003). Mammalian Ric-8A (synembryn) is a heterotrimeric Galpha protein guanine nucleotide exchange factor. *J. Biol. Chem.* **278**(10), 8356–8362.

Tandon, A., Bannykh, S., Kowalchyk, J. A., Banerjee, A., Martin, T. F., and Balch, W. E. (1998). Differential regulation of exocytosis by calcium and CAPS in semi-intact synaptosomes. *Neuron* **21**, 147–154.

Tanis, J. E., Moresco, J. J., Lindquist, R. A., and Koelle, M. R. (2008). Regulation of serotonin biosynthesis by the G proteins Galphao and Galphaq controls serotonin signaling in *Caenorhabditis elegans*. *Genetics* **178**(1), 157–169.

Taylor, S. J., Chae, H. Z., Rhee, S. G., and Exton, J. H. (1991). Activation of the beta 1 isozyme of phospholipase C by alpha subunits of the Gq class of G proteins. *Nature* **350**(6318), 516–518.

Thomas, J. H., and Robertson, H. M. (2008). The *Caenorhabditis* chemoreceptor gene families. *BMC Biol.* **6**, 42.

Tokuoka, S. M., Saiardi, A., and Nurrish, S. J. (2008). The mood stabilizer valproate inhibits both inositol- and diacylglycerol-signaling pathways in *Caenorhabditis elegans*. *Mol. Biol. Cell* **19**(5), 2241–2250.

Topham, M. K. (2006). Signaling roles of diacylglycerol kinases. *J. Cell. Biochem.* **97**(3), 474–484.

Trent, C., Tsuing, N., and Horvitz, H. R. (1983). Egg-laying defective mutants of the nematode *Caenorhabditis elegans*. *Genetics* **104**(4), 619–647.

Turrigiano, G. G., Leslie, K. R., Desai, N. S., Rutherford, L. C., and Nelson, S. B. (1998). Activity-dependent scaling of quantal amplitude in neocortical neurons. *Nature* **391**(6670), 892–896.

van der Linden, A. M., Simmer, F., Cuppen, E., and Plasterk, R. H. (2001). The G-protein beta-subunit GPB-2 in *Caenorhabditis elegans* regulates the G(o)alpha-G(q)alpha signaling network through interactions with the regulator of G-protein signaling proteins EGL-10 and EAT-16. *Genetics* **158**(1), 221–235.

van der Linden, A. M., Moorman, C., Cuppen, E., Korswagen, H. C., and Plasterk, R. H. (2003). Hyperactivation of the G12-mediated signaling pathway in *Caenorhabditis elegans* induces a developmental growth arrest via protein kinase C. *Curr. Biol.* **13**(6), 516–521.

van Swinderen, B., Saifee, O., Shebester, L., Roberson, R., Nonet, M. L., and Crowder, C. M. (1999). A neomorphic syntaxin mutation blocks volatile-anesthetic action in *Caenorhabditis elegans*. *Proc. Natl. Acad. Sci. USA* **96**(5), 2479–2484.

van Swinderen, B., Metz, L. B., Shebester, L. D., Mendel, J. E., Sternberg, P. W., and Crowder, C. M. (2001). Goalpha regulates volatile anesthetic action in *Caenorhabditis elegans*. *Genetics* **158**(2), 643–655.

Vashlishan, A. B., Madison, J. M., Dybbs, M., Bai, J., Sieburth, D., Ch'ng, Q., Tavazoie, M., and Kaplan, J. M. (2008). An RNAi screen identifies genes that regulate GABA synapses. *Neuron* **58**(3), 346–361.

Waggoner, L. E., Dickinson, K. A., Poole, D. S., Tabuse, Y., Miwa, J., and Schafer, W. R. (2000). Long-term nicotine adaptation in *Caenorhabditis elegans* involves PKC-dependent changes in nicotinic receptor abundance. *J. Neurosci.* **20**(23), 8802–8811.

Walker, R. J., Colquhoun, L., and Holden-Dye, L. (1992). Pharmacological profiles of the GABA and acetylcholine receptors from the nematode, *Ascaris suum*. *Acta Biol. Hung.* **43**, 59–68.

Wang, Y., Okamoto, M., Schmitz, F., Hofmann, K., and Sudhof, T. C. (1997). Rim is a putative Rab3 effector in regulating synaptic-vesicle fusion. *Nature* **388**(6642), 593–598.

Wang, Z. W., Saifee, O., Nonet, M. L., and Salkoff, L. (2001). SLO-1 potassium channels control quantal content of neurotransmitter release at the C. elegans neuromuscular junction. *Neuron* **32**(5), 867–881.

Watts, J. L., Phillips, E., Griffing, K. R., and Browse, J. (2003). Deficiencies in C20 polyunsaturated fatty acids cause behavioral and developmental defects in *Caenorhabditis elegans* fat-3 mutants. *Genetics* **163**(2), 581–589.

Weimer, R. M., Richmond, J. E., Davis, W. S., Hadwiger, G., Nonet, M. L., and Jorgensen, E. M. (2003). Defects in synaptic vesicle docking in unc-18 mutants. *Nat. Neurosci.* **6**(10), 1023–1030.

Weimer, R. M., Gracheva, E. O., Meyrignac, O., Miller, K. G., Richmond, J. E., and Bessereau, J. L. (2006). UNC-13 and UNC-10/rim localize synaptic vesicles to specific membrane domains. *J. Neurosci.* **26**(31), 8040–8047.

Wettschureck, N., and Offermanns, S. (2005). Mammalian G proteins and their cell type specific functions. *Physiol. Rev.* **85**(4), 1159–1204.

White, J. G., Southgate, E., Thomson, J. N., and Brenner, S. (1986). The structure of the nervous system of the nematode *Caenorhabditis elegans*. *Philos. Trans. R. Soc. Lond. Ser. B Biol. Sci.* **314**(1165), 1–340.

Wierda, K. D., Toonen, R. F., de Wit, H., Brussaard, A. B., and Verhage, M. (2007). Interdependence of PKC-dependent and PKC-independent pathways for presynaptic plasticity. *Neuron* **54**(2), 275–290.

Williams, S. L., Lutz, S., Charlie, N. K., Vettel, C., Ailion, M., Coco, C., Tesmer, J. J., Jorgensen, E. M., Wieland, T., and Miller, K. G. (2007). Trio's Rho-specific GEF domain is the missing Galpha q effector in *C. elegans*. *Genes Dev.* **21**(21), 2731–2746.

Yamada, K., Hirotsu, T., Matsuki, M., Kunitomo, H., and Iino, Y. (2009). GPC-1, a G protein {gamma}-subunit, regulates olfactory adaptation in *Caenorhabditis elegans*. *Genetics* **181**, 1347–1357.

You, Y. J., Kim, J., Cobb, M., and Avery, L. (2006). Starvation activates MAP kinase through the muscarinic acetylcholine pathway in *Caenorhabditis elegans* pharynx. *Cell Metab.* **3**(4), 237–245.

Zhang, F., Wang, L. P., Brauner, M., Liewald, J. F., Kay, K., Watzke, N., Wood, P. G., Bamberg, E., Nagel, G., Gottschalk, A., and Deisseroth, K. (2007). Multimodal fast optical interrogation of neural circuitry. *Nature* **446**, 633–639.

Zheng, Y., Brockie, P. J., Mellem, J. E., Madsen, D. M., and Maricq, A. V. (1999). Neuronal control of locomotion in *C. elegans* is modified by a dominant mutation in the GLR-1 ionotropic glutamate receptor. *Neuron* **24**(2), 347–361.

Zhou, K. M., Dong, Y. M., Ge, Q., Zhu, D., Zhou, W., Lin, X. G., Liang, T., Wu, Z. X., and Xu, T. (2007). PKA activation bypasses the requirement for UNC-31 in the docking of dense core vesicles from *C. elegans* neurons. *Neuron* **56**(4), 657–669.

Zwaal, R. R., Ahringer, J., van Luenen, H. G., Rushforth, A., Anderson, P., and Plasterk, R. H. (1996). G proteins are required for spatial orientation of early cell cleavages in *C. elegans* embryos. *Cell* **86**(4), 619–629.

Index

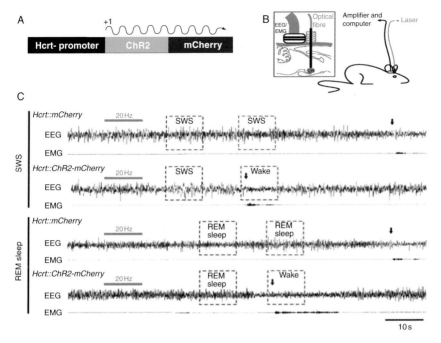

Chapter 1, Figure 1.4 (See Page 23 of this volume).

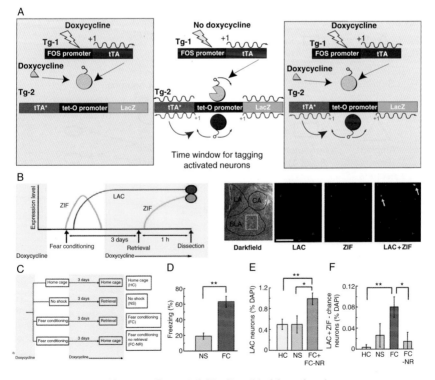

Chapter 1, Figure 1.5 (See Page 26 of this volume).

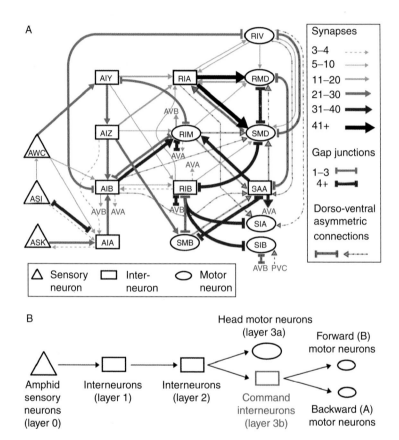

Chapter 2, Figure 2.1 (See Page 42 of this volume).

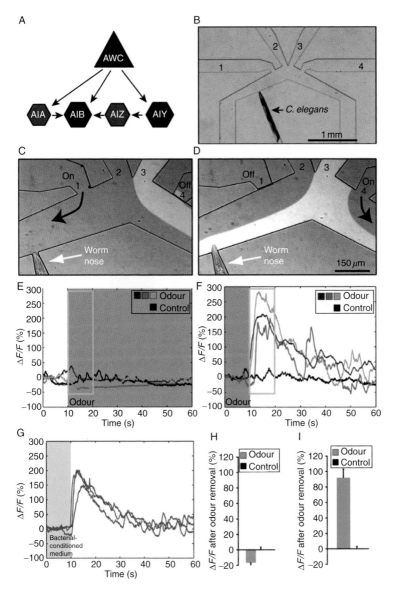

Chapter 2, Figure 2.3 (See Page 50 of this volume).

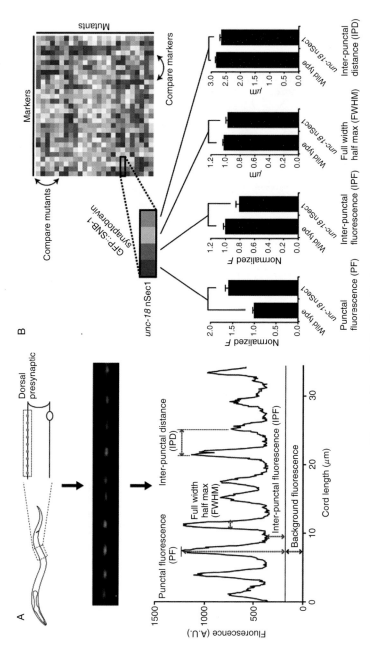

Chapter 2, Figure 2.4 (See Page 55 of this volume).

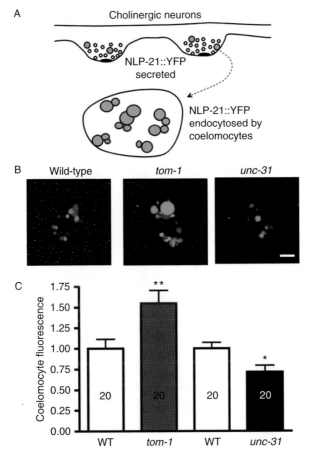

Chapter 2, Figure 2.6 (See Page 59 of this volume).

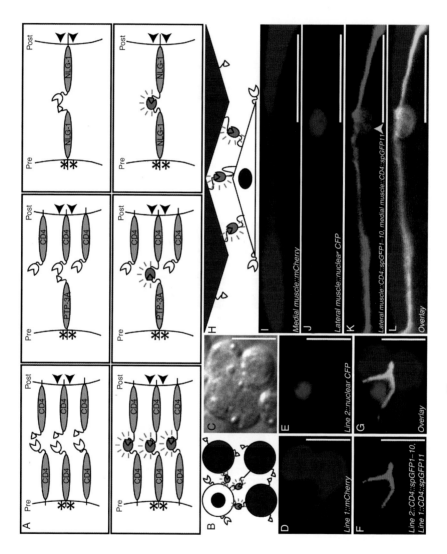

Chapter 2, Figure 2.7 (See Page 62 of this volume).

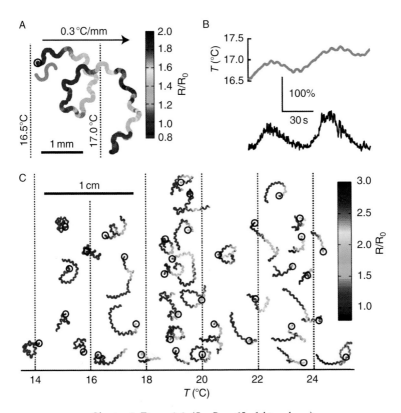

Chapter 2, Figure 2.8 (See Page 65 of this volume).

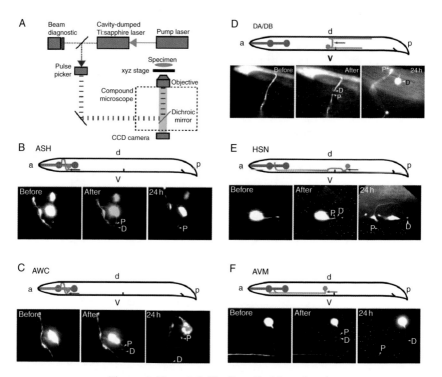

Chapter 2, Figure 2.9 (See Page 68 of this volume).